Contemporary Trends and Issues in Science Education

Volume 65

The book series Contemporary Trends and Issues in Science Education provides a forum for innovative trends and issues impacting science education. Scholarship that focuses on advancing new visions, understanding, and is at the forefront of the field is found in this series. Authoritative works based on empirical research and/or conceptual theory from disciplines including historical, philosophical, psychological and sociological traditions are represented here. Our goal is to advance the field of science education by testing and pushing the prevailing sociocultural norms about teaching, learning, research and policy. Book proposals for this series may be submitted to the Publishing Editor: Claudia Acuna E-mail: Claudia.Acuna@springer.com

Mary M. Atwater · Brittany A. Garvin
Editors

Creating and Maintaining a Home in Science Education Research

Voices Across the African Diaspora, Volume 1

Springer

Editors
Mary M. Atwater
University of Georgia
Athens, GA, USA

Brittany A. Garvin
School of Education
Indiana University
Indianapolis, IN, USA

ISSN 1878-0482 ISSN 1878-0784 (electronic)
Contemporary Trends and Issues in Science Education
ISBN 978-3-031-99381-7 ISBN 978-3-031-99382-4 (eBook)
https://doi.org/10.1007/978-3-031-99382-4

This Springer imprint is published by the registered company Springer Nature Switzerland AG
The registered company address is: Gewerbestrasse 11, 6330 Cham, Switzerland

Preface

As it is written in the song "This is Home," found in the 2008 movie "The Chronicles of Narnia: Prince Caspian," the purpose of this book is found in the song lyrics written below by Dodd, Foreman, and Watt (2008):

> Home, this is home
> Now I'm finally
> Where I belong
> Where I belong
> Yeah, this is home
> I've been searching
> For a place of my own
> Now I've found it
> Maybe this is home (this is home)
> This is home
> Now I know
> Yeah, this is home
> I've come too far
> And I won't go back
> Yeah, this is home…

Home is used throughout this book as a metaphor to represent one's professional place and space in science education. For some, home is a protected place of safety, belonging, and a sense of self. For others, home is connected to upbringing, family, history, or even national identity. This book is a collection of narratives from science education researchers across the African diaspora at different phases of their professional careers, searching for and/or reflecting on their meaning and concept of "home." This book is organized across two volumes, where a professional home includes formal and informal science education settings in which chapter authors are seeking a home, have found a home, are no longer working within a home, or have retired from a home. This book elucidates the trials and triumphs experienced by scholars of African ancestry in becoming science education researchers. We present these stories and reflections as a source of radical healing and hope where, through solidarity, (counter)narratives, and community, the next generation of science education researchers find support and courage to continue their journeys. We hope that within these narratives, readers will identify and name similar struggles, challenges, and successes that resonate with their own experiences and recount how they were able to resist and overcome. This work also serves as a critical call to action to address the shortages of science education researchers from diverse backgrounds. These (counter)narratives serve as a model for science education preparation programs for the type of support needed to foster, sustain, and retain researchers, specifically researchers of African ancestry in science education doctoral programs. This edited book will conclude with a summary of themes found across the different chapters and offer suggested strategies and tools to encourage healing and support.

Mary's Inspiration for the Book

For about 15 years, I pondered writing a manuscript about the importance and magnitude of finding a home in one's professional life. However, I never wrote the article; instead, I jotted down ideas to be included in the manuscript. That manuscript was never written or submitted. But as I neared my retirement, this concept of having a home matured in my thoughts. I now knew how to act on this idea I had had for 15 years—an edited book. This two-volume edited book is the result of a 15-year dream materialized.

Brittany's Inspiration for the Book

I was inspired to contribute to this edited book project because it offers a unique opportunity to spotlight the challenges, triumphs, and innovations encountered by those dedicated to advancing the teaching and learning of science. Through this work, readers can gain valuable insights into the dynamic field of science education, fostering a deeper understanding of the systemic and institutional challenges encountered by scholars of African ancestry and how their brilliance and persistence have transformative and restorative power.

This book serves as a bridge between researchers, educators, policymakers, and the general public. By detailing personal experiences and professional journeys, it humanizes the often technical and abstract nature of science education research. It helps readers connect with the individuals behind the data, making their contributions relevant and inspiring. These narratives also serve as a source of motivation for aspiring researchers as they showcase the diverse paths to success and highlight the interdisciplinary nature of science education research.

In closing, we wish to thank the following individuals who significantly contributed to this two-volume edited book's development: Claudia Acuna,

Karthikeyan Krishnan, Corina van der Giessen, Jayalakshmi Raju, and other unknown Springer staff who were instrumental in the publication of this two-volume edited book.

Athens, USA Mary M. Atwate
Indianapolis, USA Brittany Garvin

Acknowledgments I

Charles Hutchison
This chapter is dedicated to my dissertation chair, the late Dr. Jack Hassard, Professor Emeritus of Georgia State University, whose guidance and mentoring excavated the contours of my professional life.
Much appreciation,
Charles Hutchison

Angelia Reid-Griffin
I would like to express my profound gratitude to Dr. Mary M. Atwater and Dr. Brittany Garvin for their invaluable guidance and for providing me with the opportunity to contribute to this significant scholarship that addresses a critical gap in science education history. My sincere appreciation extends to my family, whose patience and encouragement have sustained me throughout this endeavor. I am deeply indebted to my mentors who have shaped my academic and professional growth. This work would not have been possible without the collective support of these individuals who have contributed to my academic journey in numerous ways.
Thank you again,
Angie

Sheron L. Mark
I am grateful to Drs. Michèle Foster, Mike Barnett and David Blustein for their mentorship and intellectual engagement, as well as Dr. Mary Atwater and Dr. Brittany Garvin for the opportunity to theoretically examine my academic journey.
Sheron L. Mark, Ph.D.

Acknowledgments II

Brittany's Acknowledgements for Springer Book Project

With deep appreciation and heartfelt gratitude, I first want to thank God for His divine calling on my life and for the many extraordinary Black women–grandmothers, mothers, sisters, aunties, sister-friends, and more—who have inspired, supported, and loved me throughout my journey. I am who I am because of you. Together, we embody one another's strength and resilience.

I am profoundly indebted to my family for their unwavering support as I navigate the pursuit, creation, and sustenance of a career in science education. To my husband, Christian, and my children, Gavin and Grant, thank you for sustaining me in ways I never knew I needed. Your love and encouragement have been my anchor. To my sister, Lenora, and my cousins, Edwin and Erica, your steadfast support, insightful words of wisdom, and late-night check-ins mean more to me than I can express. I am truly grateful to have you all in my life.

Lastly, I want to extend a special acknowledgement to Dr. Mary M. Atwater, who has been one of my greatest inspirations. My story is intertwined with yours, and for that, I am immensely grateful. Thank you for your exceptional leadership and mentorship—I consider myself blessed to have traveled alongside you on this journey.

Mary's Acknowledgements for Springer Book Project

I would like to thank Claudia Acuna, Editor of *Science Education, Assessment and International and Comparative Education of Springer*, for her generous support in offering a contract for Springer to publish this two-volume edited book. I also want to express my gratitude to Dana Zeitler, the Editor of Springer Contemporary Issues and Trends in Science Education Library Series, for his thorough review of the entire content and insightful recommendations to make this work more accessible and relevant to a broader audience of science educators and other fields of education.

Additionally, I want to thank Karthikeyan Krishnan, Senior Production Editor Project management of SpringerNature, for working with Brittany and me to bring *Creating and Maintaining a Home in Science Education Research: Voices Across the African Diaspora* to publication. Finally, I would like to express my appreciation to the entire Springer staff for their invaluable assistance throughout the publication process of this edited book.

Introduction to Volume 1

This two-volume edited book series is the result of a 15-year dream to capture and compile the stories, experiences, and journeys of science education researchers across the African diaspora. *Creating and Maintaining a Home in Science Education Research: Voices Across the African Diaspora, Volume 1*, features two parts, "Seeking a Home" and "Found a Home." These parts remind the reader that the concept of home is highly complex, yet important to human beings, especially when considering the following three verses in the song "This is Home," found in the 2008 movie "The Chronicles of Narnia: Prince Caspian."

> I've been searching
> For a place of my own
> Now I've found it

Home is defined in many ways, and in the field of science education, it is often very difficult for science education researchers of African descent to find a place and space that creates both a sense of belonging and success. The chapters that follow in Volume 1 explore each author's definition and journey to seek and find a home in science education. The authors are presented below in alphabetical order, alongside a brief abstract detailing their respective journeys.

Part I: Seeking a Home

Regina P. McCurdy, Ph.D.

The first chapter, written by Regina P. McCurdy, is entitled "My Journey as a Garden: Sowing, Planting, and Growing in Science Education," and discusses searching for a home. She employs the metaphor of a garden to describe her search for a home in the science education research arena. She discusses the challenges that created inner turbulence, questioning, and uncertainty as she searched for a home in the field of science education.

Heather Lavender, Ph.D.

The second chapter "Navigating to a Science Education Researcher through a Scientist and from The Fly on the Wall" by Heather Lavender shares her experiences with spaces "where others were so brazen that they did not care" about her listening ear which has informed her approach to being a science education researcher, work and role.

Part II: Found a Home

Part II includes 10 chapters; its content reminds us of the following:

> "This is home
> Now I know
> Yeah, this is home
> I've come too far
> And I won't go back
> Yeah, this is home…" (The Chronicles of Narnia: Prince Caspian, 2008)

Brenda Brand, Ph.D.

In the chapter "Drowning Out the Background Noise", Brenda Brand discovered very early that asserting her voice in determining significance was an important first step in defining the focus of her scholarship. She needed to establish her identity, which was defined by what was personally and professionally meaningful, based on the principles of social justice. She had to drown out the background noise to accomplish this.

Terrence Burgess, Ph.D.

Terrence Burgess, in his chapter, "Right where I belong: Fighting for the Space in Science that Everyone Else is Given Without Question," discusses his reality as a Black man in finding a home in Science, Technology, Engineering, and Mathematics

(STEM). He chronicles his experiences fighting for a space that others like White students and faculty members were afforded without question.

Vivien Mweene Chabalengula, Ph.D.

Vivien Mweene Chabalengula wrote an autoethnographic account entitled "Finding a Home in Science Education: Chronicle of my Crossroads Between the Zambian Sociocultural Perspectives and Formal Science Education," in which she, as a Zambian-born and diaspora science educator, shares how she navigated the crossroads between the Zambian sociocultural perspectives and formal science education. She explores the tensions and triumphs of finding a home (sense of belonging) in science education, while highlighting Zambian social and cultural practices and their relationship with science, the crossroads she encountered, and how she navigated them.

Charles Hutchison, Ph.D.

As a Ghanaian, Charles Hutchison entitled his chapter "A Home of Steel and Velvet: An African Professor's Insights for the Successful Navigation of the U.S. Academic Landscape." Here he explored his academic and research experiences, highlighting the challenges and successes encountered along the way.

Natalie King, Ph.D.

Natalie King, in the chapter "The Community is my Laboratory: Leveraging my Faith Walk to Reconceptualize Service as Scholarship in the Academy," disclosed her journey of reconceptualizing service as community-engaged scholarship so that educators, youth, and their families could be positively impacted by her presence at a university. By rejecting the scientometric index and traditional mechanisms of evaluating her impact and contributions, Natalie finds a home in science education.

Rhea Miles, Ph.D.

Rhea Miles contributed the chapter, "Fitting In: A Black Woman's Perception," which provided vignettes of her sense of belonging, persistence, identity, and resilience related to her experiences and the challenges encountered to "fit in" to a STEM-related career.

Sheron Mark, Ph.D.

Sheron Mark discussed interstitiality in her chapter, "At Home in the Interstices," and the circuitous and ongoing journey that she came to understand as she navigated the interstices of multiple disciplines and communities.

Angelia Reid-Griffin, Ph.D.

The chapter, "The 3 C's to My Success in Science Education: Connection, Communication, and Community" written by Angelia Reid-Griffin discussed how she found success and a home in teaching Science/STEM education and developed an outreach program for youth as a means of preparing the next generation of scientists, engineers, and innovators.

Melody Russell, Ph.D.

Melody Russell's chapter, "My Journey as an African American Woman at a Predominately White Institution (PWI)", related a number of experiences and people that have helped shape her into the scholar, researcher, and person that she is today. Many of her experiences are unique to the culture and climate of a predominantly White Southern institution.

Noemi Waight, Ph.D.

Using her experiences in the biking community as an example, Noemi Waight wrote in her chapter, "Found a Home: Freedom Scie-cycling," that women of color can assert their voices, effect change, find partners, and thrive in higher institutions. She discussed needing to re-center her dream, anchor joy, and forge a commitment to decolonize and strip the metaphorical spikes to chart new pathways forward.

Concluding Thoughts

The authors of the chapters in this volume articulate their ongoing challenges and struggles, particularly in their interactions with members of the dominant culture in the United States. Their primary goal is to encourage others seeking positions in science education to recognize themselves as worthy and valuable human beings, capable of becoming effective science teacher educators and researchers. Throughout their journeys and within these narratives, the chapter authors confront attempts by others to impose identities upon them (Whitbourne, 2012)—identities that they do not own and will never accept. These authors grapple with the persistent challenges faced by oppressed groups in the United States, where those in positions of hegemonic power often seek to dictate the self-perceptions of individuals from disenfranchised cultures. These dynamics influence how these individuals perceive themselves, their beliefs, behaviors, professional aspirations, and achievements. Such struggles are not merely specific to the United States; rather, they represent a broader human condition that extends globally. Across different regions, the interplay between dominant and oppressed groups—defined by factors such as culture, ethnicity, gender, language, race, social class, or religion—varies. Yet, the underlying issues of identity and self-perception remain universally relevant.

Due to their socially ascribed memberships in specific groups, individuals experience varying degrees of oppression or privilege. Typically, people do not have the option to choose or sign up for these social groups; rather, they are often assigned these identities at birth or through social norms, and they certainly do not consent to the systems of oppression and privilege that accompany them. These group memberships have a significant impact on individuals' lives and experiences.

In society, individuals occupy positions of varying status, which can be categorized as either high-status or low-status. High-status actions are characterized by a dominant and assertive posture, accompanied by verbal expressions of confidence, leadership, and authority. In contrast, low-status actions are often characterized by submissive or passive postures, with verbal messages that convey agreement, compliance, acceptance, and support (Nieto & Boyer, 2006).

When individuals from marginalized or devalued social groups achieve success and display high-status behaviors, it often provokes feelings of anger and fear among members of the dominant group. This reaction can lead to the implementation of coercive policies and practices aimed at maintaining the status quo. This two-volume book explores the coercive policies and practices that have challenged these Black authors in their pursuit of becoming science educators, highlighting the systemic barriers they encounter in their professional journeys.

References

Dodd, J., Foreman, A., Watts. A. (2008). This is Home [Recorded by Switchfoot and song by A. Watts] [Movie, 2008, The Chronicles of Narnia: Prince of Caspian (Disney).

Leticia Nieto, L., & Boyer, M. (2006). *Understanding oppression: Strategies in addressing power and privilege.* Retrieved from https://beyondinclusionbeyondempowerment.com/wp-content/uploads/2019/12/nieto-articles-understanding-oppression-2006.pdf

Whitbourne, K. (2012). *The me I know: A study of adult identity.* Springer Science & Business Media.

Contents

Editors and Contributors

About the Editors

Mary M. Atwater, Professor Emerita of the University of Georgia, is a leader in multicultural science education, author of articles and book chapters, and co-editor of edited books.

Brittany A. Garvin, Ph.D., is an Assistant Professor of Science Education in the Department of Urban Teacher Education at the Indiana University Indianapolis School of Education.

Contributors

Brenda Brand Virginia Tech, Blacksburg, VA, USA

Terrance Burgess Department of Teacher Education, Michigan State University, East Lansing, MI, USA

Vivien Mweene Chabalengula University of Virginia, Charlottesville, VA, USA

Charles B. Hutchison The University of North Carolina at Charlotte, Charlotte, NC, USA

Heather F. Lavender Syracuse University, Syracuse, NY, USA

Sheron L. Mark College of Education and Human Development, University of Louisville, Louisville, KY, USA

Regina P. McCurdy University of North Carolina Charlotte, Charlotte, NC, USA

Rhea Miles East Carolina University (ECU), Greenville, NC, USA

Angelia Reid-Griffin UNC Wilmington, Wilmington, USA

Melody L. Russell Auburn University, Auburn, AL, USA

Natalie S. King Georgia State University, Atlanta, GA, USA

Noemi Waight University at Buffalo, Buffalo, USA

Part I
Seeking a Home

Chapter 1
Navigating to a Science Education Researcher Through a Scientist and from the Fly on the Wall

Heather F. Lavender

1.1 Life in the Midwest

I always thought the thing to do was to put family first. I fell in love with a person I thought to be smarter than me. I loved talking to this person who was at the time completing a master's degree in English and one day I had an aha moment and said, "this guy just might be smarter than you." Throughout dating others in high school, my father would tell me, "That boy isn't as smart as you," and to be honest…they were not. Why is this important, because my journey in science is entangled with this guy, I thought to be smarter than me and now known as my husband. While our story precedes life in in the Midwest, the Midwest is the first place I worked as a scientist.

I prioritized my husband advancing in his career first and why did I do this? I saw us as a team of two; Team Lavender is how we would at times reference our union. My husband began his doctoral program two years ahead of our marriage at a university in the Midwest, as I was finishing my Bachelor of Science at a southeastern school. I was not entirely certain of what I wanted to do and after finishing a BS in microbiology, my grades were not the grades that got one into medical school and so I took on an entry level job in a research laboratory to help support us. When looking further into marriages of Black Americans, Black middle-class mothers, though taught to be self-reliant and "strong" place marriage in a higher position than their careers (Barnes, 2015). While I was then, not a mother and certainly, not in the middle class at this time, I saw the upward mobility of team Lavender headed in that direction.

H. F. Lavender (✉)
Syracuse University, Syracuse, NY, USA
e-mail: hflavend@syr.edu

M. M. Atwater and B. A. Garvin (eds.), *Creating and Maintaining a Home in Science Education Research*, Contemporary Trends and Issues in Science Education 65,
https://doi.org/10.1007/978-3-031-99382-4_1

1.1.1 Microbiology Research in Laboratory #1

I did not go to medical school, I did not have the grades and therefore, maybe I was not as smart as I thought, but I had a B.S. in microbiology, and I was now working in a microbiology laboratory because I loved and to this day still do love microorganisms. I have since reconciled that I was smart enough, but lacked sufficient mentoring and support (Booker & Brevard, 2017; Sinanan, 2016), particularly as a first-generation college student (Havlik et al., 2020). I vividly recall a conversation with my mother where I said, "I have found what I want to do for the rest of my life." She said, "that's a bold statement." Forming this opinion was the atmosphere that I was in. I found the very first person that I worked for to be encouraging and fostering a space where each person respected the work and ideas of each other. I realized that I did not need to be a medical doctor, I needed to be amongst those that inform the medical profession, researchers. I was then believing that research such as this was the backbone of medicine, the people in the trenches. That is where my focus should lie. I began observing the professors around me and listening to the conversations they would have while standing next to my work bench. I recall two male professors being elated in the fact that a female graduate student had cried during her general exam stating that they could not build without tearing down. I recall consoling a female student as she wept in the bathroom because she had just been reduced to crap as she explained it in her committee meeting. So, as I was enjoying the research of *Klebsiella pneumoniae*, some of the women in the program were suffering and I spent years being a cheerleader for so many women through encouraging words (Dasgupta & Stout, 2014). I thought I could do the job of mentoring women through doctoral degrees better. Things became more in focus, and I knew that whatever I was to do, I would need to direct my efforts to spaces for women. I recall a female graduate student saying that the male student in her lab was the golden boy and that she was just the lab punching bag. I believed what I would do was go to this space called graduate school and obtain a doctorate in microbiology. This would be my new focus. However, at this time my husband was four years into a doctorate degree in English. The PI of the lab encouraged me to apply to the Ph.D. program, but I thought, no that would take me years after my spouse would finish. I did not want us to live apart. Therefore, the plan was that I will join a doctoral program once he finishes and he takes on his first job, like taking turns to help each other, right? I think that should be the first thing we do, but at the same time we also knew that we would want to have a family. When would we do this? So that had to become the new first and next thing that I would do. I was on this wall in the Midwest for four years.

1.2 Returning to the Southeast

1.2.1 Microbiology Research in Laboratory #2

When we moved to a Southeastern state I worked as a Research Assistant in a microbiology laboratory that had just relocated from a Midwestern state to a Southeastern state, so they were as new to this state as I was. I secured the job via a telephone interview arranged by the microbiology department chair at my prior Midwestern university. In this new laboratory, I encountered something that I had never encountered before, a hostile work environment. I was surprised. I was shocked and at times even afraid. I called the Principal Investigator of the laboratory that I had left in the Midwestern state to ask which was the anomaly, this new space or the space that I had left. There I was a research assistant in this new laboratory, not only was I the only Black person in the lab, but I was the only United States born citizen. The Principal Investigator, Lee Chuong (pseudonym) was Chinese and his wife who was also Chinese worked as a Research Scientist, both not born in the United States. Two of the three Postdocs were also Chinese (not born in the United States), with Trunc (pseudonym), a female being the managing Postdoc. The third Postdoc was Asian, from Bangalore, India. I had never experienced individuals saying negative things towards U.S. born citizens. Up until this point in my life, I had endured racism due to being Black, but never due to being a U.S. born citizen. When a staff member of the office personnel, Mrs. Stacey (pseudonym) who handled the departmental ordering would take a day off, members of my laboratory would berate her, saying that like all Americans she was lazy, deserving of a low paid job and that not coming to work was the problem with all Americans. These things would be said during our laboratory meetings, and it was as if I was not there and simply a fly on the wall, and I knew the things being said had to be even worse when they switched to speaking Mandarin. After a year in this laboratory, another U.S. born citizen Amy (pseudonym) joined the lab. Though she was younger than me and she was not Black, I did not feel alone. Amy was trained by Trunc and Trunc continued to do things such as leave demoralizing notes on other's workbench. These notes would refer to an action as stupid or point out the time you left or arrived to the lab in reference to her coming and going or criticize the arrangement you chose for placing your tubes in the incubator. There was one occasion when we all arrived to the lab, and the freezer holding the enzymes had been wrapped shut with duct tape and no one could open it. The explanation that Trunc left on a note attached to the freezer was that an enzyme that was out of place and she did not want someone to sneak the enzyme back into place. I do not know why, but I began documenting the occasions that Trunc verbally belittled someone or left a demoralizing note. Lab members learned that I was documenting these occasions and so they began passing notes left to them to me. I also do not know why I stayed in the space for so long. The first surprise for me was during the lab meetings when lab mates yelled at one another. Now at this time I was only five years out of college. I had never witnessed a workspace of people yelling at one another. I surely had never witnessed a workspace where people were yelling at one another using a different language. There I was experiencing "culture shock" and identified as a member of the dominant group (U.S. born citizen) immersed in another culture that was likely a norm for them (Cupsa, 2018). After the lab meeting when I first witnessed this, the Principal Investigator came to me and asked what did I think of the lab meeting? I told him it was very strange and uncomfortable and that I had never been in such a space. He only said that he did not know what to do about Trunc antagonizing others this way. We were in a new building, and no one could hear the dysfunction that occurred in our lab meetings. This even happened again and again. It was no longer shocking. There was one day when the other American (U.S. born citizen) research assistant, Amy (pseudonym) had words with the leading postdoc that was the bane of everyone's existence there. Amy needed to take some items to a floor beneath ours and Trunc instructed her to include additional items. Amy stated that she would return and Trunc argued that she could carry all of the items. The exchange ended with Trunc telling Amy that she would just push her down the stairs for all of the items to get to the floor beneath us quicker. Amy considered this a threat and called the police. Because the P.I. was out of town, when the police arrived, the department chair came stomping down the hallway as well alongside the police. I gave an accurate statement of what Trunc had said to Amy. However, the other Postdocs of the lab said they heard nothing, which was untrue. There was no way that they had not heard the conversation when we were all in the same space and the dispute was so loud and now, they had become the flies on the wall and unwilling to speak up. The following week when the P.I. returned he told me and Amy that we would have to leave his lab. I pleaded that I had done nothing wrong and under no circumstance would I, a Black female lie to a police officer. I explained for over a year I witnessed Trunc belittling members of the lab, and I did nothing, why would I intentionally cause trouble now? He removed all of the projects that Amy and I were working on and reassigned us to do the same thing day after day, only switching each day with what the other was doing. It felt exactly as it was meant to feel, like torture. Within two months Amy found another job and she left. At this time, I was pregnant with my first child while taking one class and I thought that I could endure the lab until my baby was born. My Grandmother, an Obstetrics and Gynecology nurse continuously said to me, "Heather, this type of stress is not good for your baby." When I think back, with the realization of the high morbidity rate that Black mothers and babies have in childbirth (Montalmant & Ettinger, 2023), circumstances such as these add to those statistics. After Amy left, I carried all my documentation to the chairman of the department and explained what was happening in my lab. I was told that Dr. Chuong could manage his laboratory however he saw fit and if I did not like it then I should leave. I additionally, carried these concerns to the Equal Employment Opportunity Office and they found nothing wrong with the lab I was in. Then, one day the P.I. asked me to remove a small 4 × 6 print-out of the Ten Commandments that was posted at my desk, and I said no. The next day he informed me that I had been insubordinate and that he would either fire me right there on the spot or I could resign and so I resigned, and none ever knew at the time that I was pregnant. I was a fly on this wall for one and a half years.

1.2.2 Biochemistry Research in Laboratory #3

I started this lab in October with a secret. I felt awful for the secret, but what was I to do and…why did I feel bad? Who should feel bad for being pregnant, according to Percival Carter (2024), many women do. None the less, the conversation is as vivid as yesterday, when I told Alex Tackman (pseudonym), the lab P.I., "sometime in May I will need to take a leave from your lab." I guess I felt bad, because Alex was a new Assistant Professor (only two years older than me) and I knew what producing in a lab meant for someone just starting out. Having completed a two-year Postdoc in New York, NY, after finishing his Ph.D. in four years from this same institution, I was the first person that Alex hired. He did not know how to order items nor organize a lab, and I knew all of these things. We got along wonderfully. I had my first baby and returned to this lab after eight weeks of maternity leave. After two years in Alex's lab, he had not secured a grant, and the department could not extend his start up funds to support my salary and the salary of a Postdoc that he hired after me. Alex explained that he had an obligation to look after the career of a Postdoc versus me and I truly understood this.

After moving on to another lab, I would occasionally visit Alex and sometimes we would walk to get a cup of coffee. There was one point when Alex wrote an email to me, and he said why it was that he did not host a going away gathering for me when I left his lab. He said that he felt embarrassed that he had not secured a grant to cover my salary and had to let me go. He also confided that he was angry when I told him that I was pregnant, but throughout my pregnancy he felt bad for those feelings because he saw that I was, "an extraordinary research assistant in the lab all throughout my pregnancy and even when I returned." He said, "in hindsight, I wish I had let the Postdoc go and kept you." I do not know what compelled Alex to write this email. Though we occasionally, continued to grab coffee, we never spoke of that email, per his request in the email. For three and a half years, I enjoyed being a fly on this wall. Many years later when Alex received an inaugural named chair, I drove eight hours to attend in accepting his invitation and he is listed among those that I thanked in the acknowledgements of my dissertation. I was a fly on this wall for three and a half years.

1.2.3 Molecular Biology Research in Laboratory #4

The opportunity came for me to work with *Xenopus laevis*. I worked for a female Research Assistant Professor, Amy Worthington (pseudonym) and we seemed to have a great working relationship. She carefully educated me with everything that I needed to learn regarding caring for the animals. I had never harvested oocytes, nor did I know of the stages of oocytes. The oocytes had a very thin layer around them, and we had to use extremely fine tipped forceps to remove the layer without puncturing the oocyte. After removing the thin layer from stage four oocytes, we would inject the oocytes with mRNA and then harvest them at stage six to analyze temporal and spatial gene regulation. Mei Ling (pseudonym), a Filipino woman clearly older than myself and Amy, joined our lab and as I would do my work on one side of the bench and overhear Amy talking to Mei Ling, I never understood why Amy spoke so rude and mean to Mei Ling, I only knew that I was not on that receiving end. She would say things like, "it makes no sense for you to [anything not related to work] or "that sounds stupid" if they happen to be in a conversation regarding Mei's cultural practices. Amy was from the United Kingdom and once when there was a visiting graduate student in our lab, Amy said, "Americans like to raise their daughters thinking that they are princesses versus just telling them that they can do it themselves," as the graduate student showed off pictures of her niece dressed as a princess for Halloween. After some time in Amy's lab (I'm not sure how long), I informed her that I was pregnant with my second child. My husband and I had already put off attempting to conceive a second child by a year due to the uncertainty of my finding another lab to move to, but we stood firm in that we would not delay another year. Some days I would take a nap in a room that only our floor custodian used, I was so tired. In this time, Amy secured a tenure track faculty position in a western state. When Amy informed me that she would be moving, we agreed upon a time window for me to leave the lab and wrap up experiments. Two months prior to this window, Amy began receiving calls to recommend me not only for a job but also to join a graduate program in microbiology. Amy informed me that she had a problem with this, and I confusingly asked what the problem was. She stated that she needed me to remain in the lab up until she would be leaving, which was two months beyond our agreed upon time window. I said that I could not do that and face to face she told me, "Well that totally f**ks me!" For the remaining time in Amy's lab my relationship with Amy turned sour, like Mei Ling's and there I was, again under stress and fully pregnant. For a year and a half of holding to the wall and observing how rudely another was being treated, I suddenly became visible, just not in a good way. Looking back, I realized that this was about the most selfish thing for a person to do, to ask me to delay finding a job to support my family when she already had a job secured...who does that? During those final months, I would talk to Mei Ling and she was convinced that she absolutely had to remain in Amy's lab up until the moving truck would arrive. I was a fly on this wall for two years.

1.2.4 Microbiology Research in Laboratory #5

I joined Darin York's (pseudonym) lab and the research on *Escherichia coli* as a full-time graduate student to pursue a doctoral degree in microbiology. Darin was a new professor to the area two years prior to my joining his lab but after having cultivated a 24-year career in a Northwestern state. I truly felt that this was the right thing to do, but there were also times that I felt as though I was going through the motions. I was not sure if my whole heart was in it, but I absolutely loved the work of microorganisms, and I still do. I love their world and all that they are capable of. I began this work in 2009 as I was pregnant with my second child and my household was financially able for me to earn only a graduate student's stipend. I interviewed for the graduate program as I was visibly pregnant. A complexity to the lab is that I was down the hallway from the contentious lab that I worked in when I was pregnant with my 1st child. There was a piece of equipment in the contentious lab that my current lab frequently used and the person training me on this equipment took me into the contentious lab. I had no problems being in the space, but later Darin told me that the P.I. of the contentious lab informed him that I would be the only person of Darin's lab not permitted to enter his lab. Receiving this news hurt, because I thought that if I could forgive how that lab operated and treated me and others and do the work of science, why would he attempt to stop that. I could not reconcile why people continuously and intentionally aim to hurt others. Well, Darin responded to me with, "whatever you need to do this work, I will provide it and I'm sorry to have to tell you this and I'm sorry you had to go through whatever occurred in Lee Chuong's lab." For once, I did not feel totally like a fly on the wall as others passed by, even if some of my instructors were aware of my time (five years prior) in the contentious lab because this was all in the same department. Time progressed, I gave birth to my second child. My family's plan was that I would finish school and then we would look for jobs and move. We would aim to be at the same university. As my husband continued his teaching and research and advanced to Associate Professor, he held off applying to other universities as we waited for me to complete school. As I completed coursework I, occasionally conducted kitchen and outdoor science for my older son and the three other boys of my cul-de-sac and an additional boy that my husband escorted to our house when he walked our older child from school. Because I worked 45 min away from our home, my husband was always the pick-up and drop-off to and from daycare and school for our children. These group of cul-de-sac boys would sometimes ask my son, if his mom would be doing a science experiment today. I loved their excitement, the same excitement I saw when I would act as an agent of my laboratory to talk to a K-12 classroom about microbiology. Years later I came to know that their excitement was in part to how I presented the information. When I came to the junction of a qualifying exam, I did not pass, and I did not ask for an extension. It felt painful, yet I also had this pang of relief (Burford, 2017). At that point I had completed all the qualifications for a master's degree in science and I thought—it is over. I informed my husband that we were now free to leave this state and go wherever he could find a new opportunity, for we had never intended on staying here. I recall Darin telling me that you need to allow yourself a moment to grieve this lost and it felt strange that I was not feeling grief and around the corner my husband secured a job at a research one southeastern university, the same place that we had met and once lived and never thought the road would return us to. I was on this wall for three years.

1.3 Another Southeastern State

1.3.1 Microbiology Research in Laboratory #6

In 2013 I began working in an atmospheric microbiology lab, headed by Chris Brentwood (pseudonym) still at the same university that my husband was now working at. We were elated to be at the same campus. This lab focused on microorganisms that live in extreme environments of the upper atmosphere and cold glaciers and lakes beneath ice. My project consisted of collecting rain and filtering for microorganisms that produced an ice nucleating protein, a protein that mimics the structure of ice. If you've heard of crops being devastated by a microorganism, chances are the ice nucleating protein is produced.

This was my first time participating in research with samples collected beyond the four walls of a laboratory and I found it fascinating. An additional duty with this project was conducting a summer K-12 teacher's workshop. The workshop's goal was to combine the focus of our lab with an element of the teacher's curriculum. For two weeks, the teachers used our equipment, modified our experiments appropriately to their grade and school resources.

The teachers landed on constructing a weathervane and learning how to understand data from the National Oceanic Atmospheric Administration to compliment weather events, as was practice in our lab. I wanted to continue with outreach and further develop a bridge to the K-12 community, particularly after we completed our teacher's workshop. The following fall we maintained contact with the teachers and supplied materials for them to conduct an experiment related to the weathervane construction and usage. I wanted to write up what had taken place with these teachers in a manuscript and so I began looking for manuscripts involving teachers. When I came across education manuscripts, I did not understand what was meant by qualitative data. I felt the answer could be found at the university's School of Education. I found a class and by description, I thought this class could teach me how to understand qualitative work, but the class required permission for me to enroll. I made an appointment with the graduate program director and asked to be enrolled. I explained that I wanted to utilize my understanding and years of working as a scientist as a bridge to the K-12 community and this person said, "Have you considered a Ph.D. in science education?" And I thought, no sir, I just finished a master's degree a year ago in microbiology, I've had my fill of graduate classes, just one class please and I'll be on my way. He then said "I will enroll you in this class. However, you will not be at an equal footing with the students of this class because they are all in a doctoral program and they are very clear about what they are after. Might I recommend a different class for you?" I took the 'different class', loved it and then enrolled in another 'different class'. I was now open to considering a doctoral program in science education and my children were age four and eight. I spoke again with the director and Dr. Whissle (pseudonym) and there began my journey into a doctoral program in science education. After completing three classes in science education, I applied to the doctoral program and continued to work in the atmospheric microbiology lab.

I had also been asked to take part in a field season on the Matanuska glacier in Alaska. I must admit when Chris Brentwood first asked, my first thought was, what will you do with your hair?

The first thought was not what will happen in terms of the care of my children, I felt confident that my spouse could care for his children as well as I could. When I got home that day, I told my husband that Chris Brentwood had asked me to be a part of the team for an Alaskan field season and I was enthused about going and my husband said "absolutely you should go, you've never participated in research like this. This could be very exciting." I saw the opportunity as more experiences for me to share with K-12 students. I returned to lab the next day telling Chris, yes, I will go to Alaska for a field season.

In prepping to go to Alaska, Chris was not in the country. Everything that had to be done prior to was my responsibility. This was my first time being responsible for sending needed materials for a five-week field season to the project engineers. On one hand I thought this is too much and, on another hand, I thought, wow—you're trusting me with this. So much for being a fly on the wall. There's a lot of coordination involved in prepping for a field season. You ensure that the lab has double the equipment. Equipment that you will currently use in the lab and that lab mates will use and then equipment that you will also use in the field. At the time I was responsible for doing an assay that looked at ATP usage of glacial water melt. Therefore, I had to make sure that I kept accurate records of the serial and lot numbers of enzymes that I used, just in case a reaction had unexpected results. What I learned about a field season…in conjunction to a lot of science taking place—there's a fair amount of alcohol consumption.

Of course, I was the only Black person there and I had my hair braided and I didn't have to worry about taking a shower, because no one took showers. Our bathroom door was a string that crossed two bushes, and further back was our "bathroom." In "bathroom," I mean a latrine dug into the ground where we deposited our bodily excrements. I later learned that people would just urinate wherever they wanted to and used the latrine only for solid excrements. To my shock, I was the only one using the latrine for pooping and peeing. When we were at the camp, situated at the base of a mountain we used the latrine. However, on Matanuska glacier, where we hiked 45 min through the moraine of the glacier and then up to the glacier to our worksite, the "bathroom" was another issue. On the glacier you just had to squat on the glacier and collect your soiled toilet paper in a zip lock bag and be certain that you selected a spot that was down glacier from our sampling location with some level of privacy. I share these details because, when speaking to K-12 learners the one thing they wanted to know most about was how did we go to the bathroom. I wanted to let students know that nothing should stand in the way of you and wanting to do science, not even what you as a Black girl would do with your hair, because I also received that question. Given the attention to Black women and our hair (Versey, 2014), I am surprised I was only asked the question about my hair and field work once.

There were several teams from different universities, each fulfilling a specific purpose such as mass spectrometry, glaciology, computer science and engineering. In this space, not entirely a graduate student, I learned other aspects from the other teams' male principal investigators, and this became my fly on the wall moment. I will not say that the things that I learned were all negative, just different. Some spoke of the frustrations of managing graduate students. One person even said, "what was she thinking, I mean we all know how this happens" referring to a female graduate student that was pregnant. So, somewhere there was a woman graduate navigating being pregnant and graduate school and here in Alaska, her graduate advisor shaming her to others for her choice (Percival Carter, 2024). I also learned how we all had to be a part of a team. Everyone relied and depended on the information from one another. As the microbiologist, we had to inform the team of best practices, in not contaminating the cryobot before and after such that the collected samples were reliably clean samples. We relied on the project engineers to tell us the melting depth of the sample that was being collected so that we could mathematically calculate the percentage of light at that depth, therefore impacting the activity and identity of collected microorganisms.

When hiking to the glacier site, we were not walking on a defined path, but a path that was macheted out by the project engineers, therefore we walked constantly looking down when hiking between the camp site and glacier site and we had to loudly shout "HEY BEAR!" every five to ten steps. Since the moraine of a glacier has brush that is as high as the average person is tall, you cannot see a bear when it walks on all four feet. According to the day's work, different groups would be on the glacier and the rule was that at least two had to be together when going back and forth. Accompanying me became the preference of many because I shouted hey bear the loudest and in three to four step increments. They joked, "no bear's going to get Heather." I was not thinking of "those loud Black girls," (Fordham, 1993), because I did not yet know about Signithia Fordham's work, I was thinking of Hollywood movies where the only Black person is the first to die. I'm certain they never noticed that I always positioned myself in the middle of the hiking line, thinking this would help to keep me safer. Surely, there came a day when a bear was in the moraine area between those of us that were leaving the glacier and on our way to the campsite and he or she brought a friend (i.e. two bears).

Aside from work we also watched movies together in the larger community tent. We had a camp manager that would make sure our dehydrated food was prepared. Our water was collected from a stream and filtered for drinking and cooking. Each day several members collected wood for our daily fires and sometimes someone would play the guitar to familiar and unfamiliar tunes. It was culturally shocking how they all knew all the words to Hotel California, but I only knew the chorus. We concluded that field season and believe it or not I said "yes" the following year for a second field season alongside many of the same teams at the same time of the year. Unfortunately, at the time of the second field season, it was known that Chris Brentwood had signed an offer to relocate to a different Southeastern university.

I shared with Chris a few labs that I might target for working in and to some he cringed saying, "oooo I could not forgive myself if you worked for that person." I later realized that I would exit this space of laboratory bench work on a high note. I realized that each lab had been a roll of the dice, landing me in great spaces or challenging spaces and I was tired of rolling the dice. I made the decision that this lab, that I enjoyed being a part of would be my last lab and the last time that I would hold a pipet.

When I processed the last sample and left the building, it was sad and scary for me. I walked through the outside corridor underneath the arched breezeways realizing that it was time to be brave in new dreams. After 15 years of working in microbiology labs, this was the last day. Honestly, I cried because it was a lost in more than one way (Archer & Rhodes, 1993; van Eersel et al., 2021). I was saying goodbye to a part of my identity. Weeks later, I returned to the lab since the last graduate student had also finished her work in the lab and Chris returned to officially navigate the movers through packing up the remaining equipment. As the truck was being packed, I took pictures and said farewell to the best lab that I had worked in and had also brought extraordinary women in science into my life. It was time to find a home, a space that I could be for longer than three to four years. I continue to have a great relationship with Chris Brentwood and perhaps somewhere in the future we will collaborate. I was on this wall for three years.

1.3.2 Working at the Office of Strategic Initiatives

The bravest thing I think I may have done was speaking up for Dr. Pamela Genesse (pseudonym). My project was the education outreach for a consortium of mechanical engineering faculty from five of the state's colleges. We had a site visit to the National Science Foundation (NSF), meaning that we would travel to NSF to present the project's Year 1 and Year 2 milestones. Our team of 15 presenters, myself being the only female, gathered two weekends per month for four months to bring our presentation together. I was stunned when we practiced even on the day of the Super Bowl. Because our universities were dispersed across the state and everyone had teaching duties, the only time for us to gather and meet were the weekends. The schedule of presentation and questions at site visits are carefully adhered to with a timekeeper moderating the sessions of the day. We presented to a panel consisting of two women and three men from various universities. On the last day, the last word said to our team came from one of the women and she said, "the next time we see your team we hope to see a better representation of gender diversity." It was as if she had held this comment for precisely that moment because time ended, and we could not respond. I'm certain that if we could have responded we likely would not have divulged that during one of our practice sessions we considered a team member who was a female engineer and married to one of the male faculty engineers, but he said that someone had to stay at home with their children. At that moment several turned their gaze towards me (the only female) as I held this disdaining look upon my face. Had he forgotten that I was there or had he simply not cared. Flies on the wall do not buzz about, they simply stay unnoticed. I think a few others directing their gaze to me, prompted him to further explain his sexist comment. Yet, I'm certain that was not the response that this female panelist would have been satisfied with. We returned home feeling that we successfully presented the way in which we had completed projected milestones, and we would receive the remainder of our project's funds. The following year, the College of Engineering hired a female engineer Assistant Professor, Dr. Pamela Genesse and she joined our team. Her joining our team was just in time for our reverse site visit which meant that NSF panel members would come to us, tour our facility and check on our progress post our site visit. The schedule of presenters was set by the senior engineering faculty and Dr. Genesse was placed to present a portion of the education component with me. In a meeting with Dr. Genesse as I was explaining to her what she would present, I realized that this was not for her to present. I pondered why was there another Assistant Professor (male) presenting his actual research during the research portion of the presentation. These ponderings led me to send a group email to all the engineering faculty on the team and recapturing what was said to our team at the site visit just as time expired and we could not give any type of rebuttal. I informed them that Dr. Genesse did not belong in presenting in the education segment and to place her, the only female engineer in a section with the education coordinator and not presenting her research while others would be presenting their research was unfair and disrespectful to the work that she had accomplished. I recall pressing send on my keyboard and how nervous I felt, even though I had been with this group of people for the past three years. I wasn't exactly one of them, I was a bit of the outsider, the fly on the wall again and the casual conversations two to three would have as we all sat at a table sometimes reinforced that I was the outsider. During those conversations, if a student and the judgement of that student was part of a conversation then so be it, I did not know these students and I did not insert myself into the conversation. I had never been in their faculty meetings and so the contempt for whatever was being mentioned and enforced because of their faculty meetings could be discussed without regard for my listening ear. None the less, on this day, concerning Dr. Genesse, I inserted myself. I felt nervous when I sent that email in what I considered defending Dr. Genesse. I told another person down the hallway that I'm surely going to get fired for chastising these male engineering faculty and this person assured me that I was overreacting. Well, the next day I received an email from the chair of the mechanical engineering department, and he thanked me for my concerns and said that they would revisit the presentation. Dr. Genesse was moved out of the education section and slotted into a space with other engineers, and she presented her research. Being a graduate student in education and reading manuscripts related to women and Black women in science had given me a voice and a little bit of confidence. I was enjoying my graduate educational experience and even more so enjoying the familiarity and comfort of this Southeastern school.

I am a graduate of this school. My spouse and I love this school, because it holds the beginning story of us. My children loved living in this region of the country and the culture of the space. We bought a dream house there…yet a point came for us to seriously consider leaving. My spouse was once told when he arrived here that his merit was all that counted in going from Associate to full Professor, yet when he mentioned to his chair that he was wanting to go up for full Professor due to the scholarship that he had produced, his chair later responded with, "your scholarship is good, but you have not served the time as an Associate." These things are marked as moving the goal post (Baez, 2000) during the game and leaves a bitter taste. It is difficult to feel valued in your workspace when this happens. So, in being contacted by another research one Southeastern university to apply for an Endowed Chaired position on top of being offered the position, my words to my husband were, "this university that we love does not get to keep a diamond and treat it like topaz," therefore, we left. We did not really give the school an opportunity to counter. My husband kept his department in the loop throughout the entire process and we thought, "they're going to swoop in and stop this." When my husband finally secured the offer, his chair said this is a fabulous offer, we're going to be competitive and proceeded in telling my husband about his [the chair] career trajectory into administrative work and how he could see my husband as mentor to junior faculty along with other pieces of administrative work. Though my husband came to his chair saying, "we love living here, help us stay," his chair did not see what my husband needed. In that moment, my husband said, "I knew at that moment we would be leaving." I had already spoken with my graduate advisor a month ago and at the end of Spring, I would have completed my course work and my qualifying exam. My graduate advisor said your family should not pass on this opportunity and that I could collect data for my dissertation from anywhere. I was on this wall for three years.

1.4 One More Southeastern State

1.4.1 More of the Southeast and a Research One Institution

We relocated to yet another Southeastern state two years before I finished my doctoral work. My husband was taking what he considered a dream job at the state's flagship school, an Endowed Chair Professorship. We would be able to budget his new salary without my working while finishing out my doctoral work. My job was to stay focused and finish my Ph.D. I helped my family, 13- and 9-year-old boys get acclimated to the new space. We rented a three-bedroom fully furnished house from the university. Yes, it was a bit smaller than the space we left in the "dream house" of the previous state, but it was in walking distance to some cool town eat spots and my 9-year-old for the first time was able to walk and ride his bike to school. My husband would sometimes walk him to school and then catch the university shuttle into campus. He could even if he so desired catch the university shuttle from our house to campus. I knew the building on campus that was the College of Education, and I thought, I will work in that building one day. The people in that building will be my colleagues. In moving to this state, I had disrupted an established K-12 network in the previous state and it was in that K-12 network that I would have designed my proposal around to carry out my doctoral data collection. So, now I would need to rebuild a new network. My plan in doing so was to visit and get to know the educational climate and tone of the city. I learned that the schools needed substitute teachers, therefore I signed up to be a substitute teacher and taught as much as I could in the K-12 schools, three to four times a week. After about five schools, I quickly realized which schools would be ideal for the research questions of my dissertation. However, I realized other things about my new city. I realized the disparity and inequity from school to school and this realization pushed me to continue in substitute teaching in the elementary schools and even in some of the middle and high schools. Ultimately, I substitute taught at both of the city's high schools, all of the city's middle schools, and 11 of the 14 elementary schools. I would substitute teach according to the writing and word count of my doctoral work. I deemed the money I made from substitute teaching, Chick-fil-A money for my family. I laughed in saying that all we can do with it is go to Chick-fil-A. The Covid-19 pandemic abrupted my substitute in-class teaching. During the summer of 2020, I took on the opportunity to assist in writing a chapter for National Academies of Science and Engineering Medicine. My mentor for the project, a Research Scientist at a non-profit based out of Boston, Ma asked, "Heather what is your plan B if schools do not reconvene in the Fall of 2020?" I recall telling her that I had no plan B and that schools would have to convene, they had to. In working in a lab, I always had plans B, C, and D yet, her question shook me. However, as the summer crept by, I realized that I had to maneuver a new space to conduct my doctoral work. I reached out to a local private school and gained permission to conduct my doctoral work. I consulted my graduate advisor and told her of my plan. She was not initially on board because she thought my work was best in a public school, but I realized and saw the relevance of my questions as they pertained to young Black girls even in a private school setting. Over the next 1 ½ years I successfully collected data, wrote my dissertation and completed my Ph.D. I resumed substitute teaching in the Fall of 2021 sporadically as schools attempted to reopen due to Covid-19. In substitute teaching, I had once again built a network and returned my sites to working in the building that housed the university's College of Education. I was enamored by the thought that two female giants in the field of science education also worked in that building. I imagined what could one learn having those two as colleagues. Freshly, finishing my Ph.D. from a rival Southeastern school, I applied to a position of Assistant Professor, but I did not even make the Zoom list. I was not surprised, I had not been one to finish my doctoral work with many publications in education and here at this high research intensity university, the depth of 15 years of work and publications in science did not seem to hold up. In the Spring of 2022, a friend and graduate student passed along to me the call for an NSF Postdoctoral Fellowship. I read the request for proposals and thought…I can write this! By God, I CAN WRITE THIS! I would need a sponsoring researcher and who better to ask than one of those giants, and thankfully this woman, Annabelle Massey (pseudonym) was no stranger to me. I had met her two years ago at a conference back when this university originally contacted my husband, and she provided sagely advice for me. There was something about her that I recognized. Her directness felt like a soothing that I had known all my life, like my aunties sitting around my mother's kitchen table. Tracey Lewis-Giggetts (2022) tells of a Black woman in church telling a younger woman that her slip was showing and the younger woman receiving it as an older generation looking out for the well-being of this younger woman and this comes to mind in what I recognized in Annabelle. I asked if she would sponsor my NSF proposal, and she asked for me to present what I would write about, and it felt like defending my dissertation all over again. She said yes to being my sponsor and that she would begin the wheels of the university to assist and that I should begin writing. Over the next four weeks, I dropped nearly all in my life and wrote my first proposal. The personnel staff that assisted with creating a budget told me that he normally would tell people that if a request for a proposal is two to three months out and you are not finished in writing to pass it by. I told him that I think people with jobs can do that, but when you do not have a job and bounded locally, you go for it. The proposal was due on Mardi Gras day at 5:00 pm and I turned it in at 4:40 pm. Little did I know that at 2:00 pm, I was not affiliated in NSF with any university and this reset would take two hours for tech support to assign me as a non-affiliate. Long-story short, I got the grant, and I felt as though I was on cloud nine because I would be able to carry out my research and I would be in the university's education building…I was doing it.

I was so excited to begin the work of my Postdoc, but entering the space of my department was not exactly what I thought it would be. I did not have a space to work until three months after NSF had directed money to the university and I shared a back-to-back office with a graduate student. I wanted to get into the routine of coming to campus. After shifting of faculty due to an additional open space, I received my own private office. However, within a month of moving into the office, the chair of the department informed me that he might need to move me out of the office stating that he had a spacing problem and proceeded in asking me how should he handle the spacing problem. I provided an answer for him along in telling him that such an action was surely out of the parameters of a Postdoc. I was also informed that another professor's Postdoc would be coming into the office space. I asked how could I have confidential meetings in compliance with my Institutional Review Board procedure between myself and the Research Graduate Assistant that would be on my team if I shared the space. Annabelle reminded the chair of the Facility Service Agreement within my grant between NSF and the university which stated that I should have a private office. I not only remained in the office, but the talk of inserting another Postdoc into the space also ended. My existence in the department, in the building that I had once romanticized felt like a fight at every turn. My Postdoc would only be for two years and while I had desires of a position that would keep me there longer, I knew not to decorate my office and make it feel as if I had found a home, though others in the department in temporary positions had decorated and even custom painted their office. I witnessed the administrative staff helping one person move in and decorate her office. Even ordering items for my project and being reimbursed for travel was a feat. I could see what was happening—I was an outsider, I was a different type of fly on the wall. When it was time for me to carry out my first data collection the department chair was refusing to sign off on the paperwork for petty cash to my participants until I was able to inform the office personnel of the procedure. While attempting to attend a conference I had to plea back and forth with him via email that that was not part of my role, but the role of the business manager and again Annabelle had to advocate for me. In teaching a course as outlined in the solicitation of my fellowship, the Science Education Coordinator assigned an online asynchronous course to me while assigning all three sections of a course that would have been more beneficial for my professional development to her former doctoral student (who yes, had a nicely, lived in pretty in pink painted office). I thought…wait a moment, you cannot allow me to teach at least one of those sections. I nearly begged the office staff to help me in ordering items for my research. No…it was not a picnic, I was certainly a fly on the wall. Though the college was named for the first Black woman to graduate from the university, it was twisted on the inside. In the second year of my Postdoc, my husband and I engaged the services of the university's dual career hiring. We naturally were asking that I be considered for a tenure track Assistant Professor position in the same program of my Postdoc. My husband's college would pay a third of the salary, the Provost's office another third and my college the remaining third, for two years and then in the third year my college would be responsible for the entire salary. The dean of my college sent the topic to the science education program and throughout all the Fall of 2023, they rationalized the points of my Curriculum Vitae were not competitive enough for their program. They stated that I should take on a Research Scientist position in their program to see how I would develop, and I would be responsible for 60% of my salary in the third year. I explained to one faculty person in the program, that they are doing precisely what has often been done to so many Black people in academia…here, jump one more hoop. The department as a whole carried two other disciplines and when the other two thirds of the department questioned the program that I was affiliated with in citing external funding I secured from federal and city grants, my lengthy publication in science, my recent publication in science education, the participants of my research as young Black girls, and my established work of four years in the community, they were met with silence. Unfortunately, my sponsoring researcher, Annabelle was on medical leave in the fall of my second year and unable to engage in any kind of university business, including advocating for me to her colleagues and even if she could have, I do not believe the outcome would have been any different. In fact, after faculty members in the other two programs of the department questioned the faculty of the program that I was affiliated with, the faculty in my program affiliation removed the possibility of a Research Scientist position. I was no longer a fly on the wall, my research was now run over road carcass. I had a faculty member of my affiliated program tell me that they questioned if I would be competitive on the open job market; as I was on the job market and had applied to five research one universities, advanced in four of the five and finalist in two of those four. Though months of tears have simmered to a brief paragraph here, it was extremely painful for me and my family. I thought my home could have been there. My husband and I felt as though we were literally fighting for our family's survival in a town that we had come to enjoy living in. My husband's college did offer a Research Scientist position to me, where I would still be responsible for securing external funds for 60% of my salary in year three and the position was also a $10 K pay cut. However, I chose to accept a tenure track Assistant Professor position at a private school in the Northeastern area of the country. I lived in this Southeastern state for five years, and I was on the wall of my Postdoc for two years.

1.5 Life in the Northeast

As I take post on the wall in the position of an Assistant Professor, I look back and it feels like so many of my entry points into pivoting life milestones have been this sort of rough and bumpy landing, like the pilot of the comedic movie Airplane as he's attempting to land the plane, the run-way lights go off, the co-pilot pulls the gear completely out of the console, and the oxygen mask drop down, but I am here. I beat out a lot of people to secure a tenure track position at a research one institution and I also got beat out at two others. The job market of academia is a grueling process, not for the faint of heart. My husband and I are elated that I secured this position, and he is even more elated that it is HIS home state! However, the turbulence is that there was not a position for him now. We are enduring the infamous "two-body" problem of academia. It feels frustrating when I see it working for so many others, but they have likely had turbulence as well. Thus far, my space feels like sunshine. My colleagues seem to be helpful to one another and to me. Yes, I am once again building a network. I honestly consider this building capacity to be my superpower. I was ready and willing to give my effervescent self, my joy, to a place that was not deserving of it. I am at a space that seems deserving of my talents. It is not that I am worthy of them, but are they worthy of me. Perhaps I am off the wall, and this scientist turned science education researcher has found a home.

References

Archer, J., & Rhodes, V. (1993). The grief process and job loss: A cross-sectional study. *British Journal of Psychology, 84*(3), 395–410.

Barnes, R. J. D. (2015). *Raising the race: Black career women redefine marriage, motherhood, and community*. Rutgers University Press.

Baez, B. (2000). Race-related service and faculty of color: Conceptualizing critical agency in academe. *Higher Education, 39*(3), 363–391.

Booker, K., & Brevard, E. (2017). Why mentoring matters: African-American students and the transition to college. *The Mentor: An Academic Advising Journal, 19*, 1–9.

Burford, J. (2017). Not writing, and giving 'zero-f** ks' about it: Queer (y) ing doctoral 'failure'. *Discourse: Studies in the Cultural Politics of Education, 38*(4), 473–484.

Cupsa, I. (2018). Culture shock and identity. *Transactional Analysis Journal, 48*(2), 181–191.

Dasgupta, N., & Stout, J. G. (2014). Girls and women in science, technology, engineering, and mathematics: STEMing the tide and broadening participation in STEM careers. *Policy Insights from the Behavioral and Brain Sciences, 1*(1), 21–29.

Fordham, S. (1993). "Those loud Black girls": (Black) women, silence, and gender "passing" in the academy. *Anthropology & Education Quarterly, 24*(1), 3–32.

Havlik, S., Pulliam, N., Malott, K., & Steen, S. (2020). Strengths and struggles: First-generation college-goers persisting at one predominantly white institution. *Journal of College Student Retention: Research, Theory & Practice, 22*(1), 118–140.

Lewis-Giggetts, T. M. L. (2022). *Black joy: Stories of resistance, resilience, and restoration.* Simon and Schuster.

Montalmant, K. E., & Ettinger, A. K. (2023). The racial disparities in maternal mortality and impact of structural racism and implicit racial bias on pregnant Black women: a review of the literature. *Journal of Racial and Ethnic Health Disparities,* 1–20.

Percival Carter, E. (2024). Power and the perception of pregnancy in the academy. *Gender, Work & Organization, 31*(5), 1951–1975.

van Eersel, J. H., Taris, T. W., & Boelen, P. A. (2021). Grief reactions, depression, and anxiety following job loss: Patterns and correlates. *European Journal of Psychotraumatology, 12*(1), 1905348.

Versey, H. S. (2014). Centering perspectives on Black women, hair politics, and physical activity. *American Journal of Public Health, 104*(5), 810–815.

Sinanan, A. (2016). The value and necessity of mentoring African American college students at PWI's. *Journal of Pan African Studies, 9*(8), 155–166.

Chapter 2
My Journey as a Garden: Sowing, Planting, and Growing in Science Education

Regina P. McCurdy

2.1 My Roots in Fertile Soil

> That person is like a tree planted by streams of water, which yields its fruit in season and whose leaf does not wither—whatever they do prospers. (Psalm 1:3, NIV Translation of the Bible)

This garden metaphor began when I was young. I have vivid memories of my father's garden in the backyard of our family home in South Florida. He planted collard greens, tomatoes, black-eyed peas, string beans, broccoli, and other staple crops that grew abundantly in the tropical climate. He would spend hours checking his plants, monitoring them for critters and insects, tilling the ground, and watering the ground. Year after year, my mother, brothers, and I were fed by my father's flourishing, nutrient-rich garden. My mother's father also had a garden specializing in sweet potatoes and okra cherishing his time with the land and the vitamin-packed sustenance these crops contributed to the nourishment of his family.

While I have always admired people, who gardened and tended to plants and crops, it took the 2020 global pandemic for me to put hands to the shovel, to ground, to the soil for me to plant my own garden as an adult, a wife, and a mother. I spent several days removing overgrown grass and roots around the base of a mango tree to unveil the dirt below. After expending so much energy, sweat, and pain from achy muscles that had apparently been lying dormant, the ground was ready. I researched which herbs and plants were compatible, which needed sunlight, and which would be happy in the shade. I bought and replanted some plants, and I started others from a very tiny seed. I drew out a blueprint of where to plant them. I spaced them out

R. P. McCurdy (✉)
University of North Carolina Charlotte, Charlotte, NC, USA
e-mail: rmccurd1@charlotte.edu

M. M. Atwater and B. A. Garvin (eds.), *Creating and Maintaining a Home in Science Education Research*, Contemporary Trends and Issues in Science Education 65,
https://doi.org/10.1007/978-3-031-99382-4_2

Fig. 2.1 Starting my first garden

accordingly and stepped back with pride at this new journey into gardening I had finally taken as Fig. 2.1 shows.

Every day, I would check on my garden. Watering it under the scorching Florida sun. Days would lead to weeks, and I realized that as plants grew and stretched out their green limbs, I noticed other plants doing the same. These new plants were not always the ones I planted, but they, too, were growing and sometimes, crowding out the crops I was eager to see flourish. This informed me that the soil was fertile. Hidden beneath the ground's surface, the roots of each plant were intermingling, taking up water from the soil and releasing other nutrients in return. This botanical community underground was active. The components for growth were embedded into the dirt, the life of the soil of this garden. What I sowed and planted continued to grow because of the soil's fertility and the attention I gave to this garden.

When my husband, two children, our dog Sabra, and I left Florida so that I could pursue my current position as an assistant professor in science education, I knew that my garden would not be able to go with me. But undoubtedly, that little plot of land that I tended for a period had the potential to thrive for the next family due to the work I put into it. So, has been my journey in making science education home. It is in the memory and legacy of my late mother, Patricia Ann Wright Murray—the best science teacher, prayer warrior, encourager, and lover of all people—that I am able to share my journey with readers of this chapter, as she was and will always be a beautiful pink rose in my garden.

2.2 It Started with a Seed: Myself as a Science Learner

> Trees are affected by their sapling days as much as people are by their childhoods.
> —Robin Kimmerer, Braiding Sweetgrass, (2013, p. 143)

I entered into teaching somewhat reluctantly. Aside from my mother being a high school biology teacher, I had a natural inclination towards science at an early age. As a little girl, I wanted to become a doctor when I grew up, a pediatrician in particular. I am not exactly sure of the genesis of this idea; I just remember having an affinity towards wanting to treat people's bodies and ailments. The science behind how the body and other living things worked and functioned was so intriguing to me. As a teenager and young adult, I remember religiously watching television shows like St. Elsewhere, ER, and others that depicted what goes on in a hospital setting between doctors, nurses, and their patients. I enjoyed hearing about the symptoms patients had and how the medical team determined how to treat them.

I was the type of student who did not have to ask their parents for help with their science fair project. I knew what I wanted to examine, set up the experiments myself, and diligently recorded the data to analyze without much input from my mother. I remember growing cowpea plants in my backyard and testing how their growth affected the alkalinity of different types of soil, shown in Fig. 2.2. For another project, I surveyed my classmates and had some of them track their study habits and their grades as a result of exercising before studying. In my early high school years, I recall presenting what I learned about soap and teaching my classmates about the polar and nonpolar compositions of soap that help remove dirt from things we wash.

Science was in my blood, so to speak. I am sure my mother's profession in science teaching, and the memories she had of pursuing phlebotomy and medicine played a large role in my tendency to enjoy and learn about life sciences, but—while I cannot

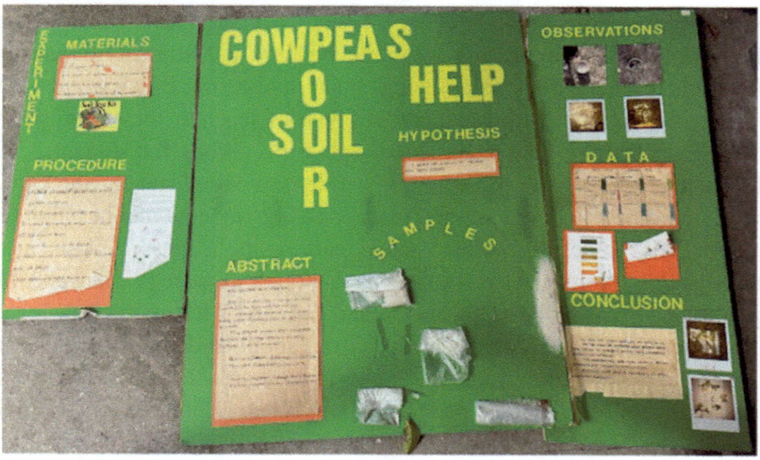

Fig. 2.2 Science fair project-cowpea growth in different types of soil

lay my hands on it—I feel as though there are other sources of my science interest that are a mystery to me. What I know now is that I enjoyed, and still do, the puzzle that science presents. I am fascinated by how scientific processes work, how professional scientists research, think, and test variables, all in an effort to put pieces together to better explain how the natural world around us works. Nevertheless, my affinity to science was probably equally matched by my desire to learn. The classroom was an environment in which I thrived, and several of my teachers played key roles in the steps towards my current work in science education.

2.3 Sowers in the Garden: My Teachers

Every well-tended garden needs to have those who place the seed into the soil, the Sowers, and Planters, those spend more time tending to the seedling as it sprouts, pokes it head up through the soil, and reaches its body upward towards the sun, while extending its roots deeper and deeper into the earth. I have been blessed to have several wonderful Sowers and Gardeners throughout my science education journey. Many of my teachers were Sowers; they placed down seeds during a season, and though they may not have known the influence they had at the time, to me, they are individuals who I have valued throughout my life. Here are a few Sowers, represented by pseudonyms, that have formed me through the years.

I attended the only public elementary (primary) school in the area from K-4th grade. I have memories of a few of my teachers in elementary school. Mr. Kelley (pseudonym) was my 2nd grade teacher. I remember him because he was the first teacher I had who incorporated computer technology into his classroom instruction. He designed his class such that when we entered the room at the start of the day, we students were able to type in and select our name in a program to mark our attendance. In this way, he gave us eight- and nine-year-old some ownership and accountability for participating in class. However, the two teachers I recall the most are Mrs. Maple (pseudonym) (my 1st grade teacher) and Miss Style (pseudonym) (my 4th grade teacher). They were strong, smart, and caring Black women. I would label them as being the "warm demander" type of teacher: they were passionate about students' learning and maintained high expectations for students' success. Mrs. Maple was smart, firm, and strong in her demeanor and stance. She knew what she was doing, knew her content, and made sure her students met her high level of expectation. Miss Style incorporated active learning in the class, took us outside of the classroom to explore, and she spoke to us as though we were brilliant and intuitive young thinkers. She was beautiful inside and out, and I felt important, loved, and at home in her classroom. They were both strict about students' behavior, but they never hesitated to show care, concern, or laughter towards these first grade or fourth grade children they loved and helped to develop. When I think of them, even though Mrs. Maple whacked a ruler across my hand for something I did (though I have no idea what it could have been!), I feel a proud connection, a kinship of sorts, with them and what they represented to me, though I was probably unaware at that time.

When I think about the science teachers that are most memorable to me, names like Mr. Green, Mr. Corey, and Mrs. Cherry (pseudonyms) all come to mind. They were all White teachers, but they exuded knowledge about their subjects and a genuine passion for teaching their students that were encouraging and motivating at the same time. Mr. Green was a soft-spoken man who taught Life Science to us seventh-grade students at a private Christian school I attended from fifth to ninth grade. No student ever felt like they could not learn and succeed in his class; his teaching made sense, and he spoke to us with a sense of hope or joy behind his slow, deliberate way of talking. Mr. Corey was a more direct communicator with a subdued and dry sense of humor. He was my anatomy and physiology teacher in high school, and while he had high expectations, I felt supported and encouraged to excel. I also felt as though he was always rooting for me to do well, and I recall him building my confidence and acknowledging my achievements in his class. Mrs. Cherry, my high-school chemistry teacher, on the other hand, was loud and had a distinctive nasal laugh that you could hear clear down the hall. She knew all of the extracurricular activities in which her students participated and was one who emanated school spirit. She made those stoichiometry problems seem like child's play. Her fast speech and explanations of her calculations on the whiteboard moved too fast for me to comprehend. However, she was very supportive of her students' learning as well, and took the time to walk through problems we had individually, during lunchtime or after school if needed. She helped me to think that I could not only conquer chemistry, but enjoy it too. Years later as an organic chemistry student in college, I was even more thankful to have had Mrs. Cherry as my high school chemistry teacher. While she was not a science teacher, Ms. Red (pseudonym), my AP high school English teacher, recognized my gift for writing. She provided encouragement and critical feedback on my essays and papers. There were moments when I felt Ms. Red could see something great in my future that I wasn't quite able to see.

Lastly, there was Mr. Fraser (pseudonym). Mr. Fraser was my high school physics teacher. He was a Black man and hailed from Louisiana. Students in our physics class easily took advantage of his mild and quiet demeanor, chatting away during lectures and venturing on their own exploits without always following the lab procedures or leaving the classroom on extended bathroom breaks. He seemed to be unfazed by it. I did not understand why he was not stricter or why he did not raise his voice. He knew physics and when he helped students at their table groups, he was patient and repeated the directions in ways that made sense to us as juniors and seniors. I was always intrigued by his expertise, and the unspoken purpose that I sensed he had for being a classroom science teacher. Fast forward a couple of years when I encountered Mr. Fraser again. I was taking an undergraduate science education course, and he was now pursuing his Ph.D. in science education. He was the graduate teaching assistant for the class. The excitement he exuded seeing one of his former high school students in his course showed on his face and in his attentiveness toward my success. He offered to call him by his first name, telling me that I was now his peer. I could never do that as that went counter to my upbringing of respecting my elders, especially those who were my teachers. But it felt great to have Mr. Fraser acknowledge me and my potential. I did not know then that I would be in this same position one day, teaching undergraduates education students about science inquiry and practices while working on my own Ph.D. Each of these teachers planted seeds of success that embodied potential I did not know I had at the time.

2.4 Accepting the Call to Teach

God cares about you and the work you do. It has meaning…
In the lab,
In the classroom,
In the field,
In the networking,
In the research,
In the writing,
In the teaching,
In the speaking,
In the testing…
In the quiet.
This work is not in vain.
This is the place where God wants you to be.

I wrote this poem as I was considering the gravity of what it means to be a person of faith in the academy and field of science. It is a reminder to me that this journey was not and is not in vain. I believe my work, this work, is a cherished gift God gave for me to humbly receive, graciously respect, and continuously refine and share whenever and wherever I am. This poem also hints at the challenges that need to take place behind the scenes, underground in the dark soil below the earth. The place where the seeds begin to find their way.

My germination process into science education took some time and struggle, as many significant biological processes do. I did not have the typical preservice teaching internship experience. I transitioned from being a biology major to science education major and then back to being a biology major during the five years of my undergraduate studies. In retrospect, I was attempting to ignore the pull and tug of teaching, because both of my parents were in education. I wanted no part of that profession, or so I thought. While in college, my father would always tell me to take education courses so when I was home on winter or spring break, I could substitute teach. I think…no, I know that my father and mother always knew I would end up teaching. I ended up substitute teaching, but not until I exhausted every hope of entering the medical field with the goal of becoming a doctor.

I became a Certified Pharmacy Technician in college and a Certified Nurse Assistant after graduating with a Bachelor of Science Degree in Biology. My perplexed father, assistant principal of a predominantly Black high school, could not understand why I was not striving for more…for teaching or school leadership. My plans to pursue medicine were thwarted by average attempts at the Medical College

Admission Test (MCAT) and medical school applications. But there was a memory I continue to remember. During an MCAT study session, I found myself explaining a chemical process from our review books to another peer in the class. While I'm sure others were around, I was confident in what I had been studying and pouring over to prepare for the practice test and shared how I was making sense of the reaction. After I finished, the classmate said something to the effect, "You explained that to me so well! I really understand it now." To this day, this memory brings me joy, because I think this was the first time that I could pinpoint being excited about the prospect of teaching science to others. It became clear to me shortly after this moment that I was called to teach, more specifically, to be a science teacher.

My teaching experience began with substitute teaching. Substitute teaching was critical in shaping me as a future teacher. I gained experience teaching different subject areas, different middle and high school grades, and students from different socioeconomic and ethnically diverse backgrounds. My exposure to these various learning environments gave me very practical tools and techniques for explaining information, monitoring students' attentiveness and participation, and recognizing the discrepancy between privileged and underserved student populations.

I took my first full-time high school science teaching position not too long after this event. As a young adult with a biology degree and only two education courses under my belt, I was not necessarily prepared for all that a teacher is accountable for in the public-school environment. Students are in your classroom to learn the content of science, that is the focus…right? What I did not know then, as a seedling science teacher, but I am greatly aware of now is that teaching science encompasses much more than content knowledge.

When I first started teaching high school science, my instruction and assumptions about teaching resembled the banking model, Paolo Freire (1970) spoke out against in his work *Pedagogy of the Oppressed*:

> Narration (with the teacher as narrator) leads the students to memorize mechanically the narrated content. Worse yet, it turns them into "containers," into "receptacles" to be "filled" by the teacher. The more completely she fills the receptacles, the better a teacher she is. The more meekly the receptacles permit themselves to be filled, the better students they are. (Freire, 1970, p. 163)

I imitated how I had was taught as a K-12 and college student, using primarily lecture-based, note taking, laboratory activities, and some group projects. Students sat in rows facing the chalkboard (or whiteboard) unless they were working through a laboratory experiment. They were expected to write down the notes I provided on the board or from the Power Point slides as I read them. I welcomed students' questions and provided answers to their inquiries, though if other students had a response to a classmate's question, I would typically allow them to respond as well. I praised students who answered questions correctly; for those who answered incorrectly, I praised their effort with "good try" and sometimes call on another student with a raised hand to respond. In my first years of teaching, if students performed poorly on the class work or tests, I would assume that they were not studying hard enough, were not taking the content seriously, or that they did not value education, thereby denying themselves a chance of attaining a good degree-required job. If their grades in the class were low, they just were not prepared, because I thought I had taught and covered everything very well. Students who were off-task or playing around in class were reprimanded, as they were a disruption and just did not want to learn. School was the means of "maintaining rather than altering structures of society" (Longstreet & Shane, 1993, p. 112). Encounters with students after class time started to needle their way into my thoughts that have never left. One student told me that he slept in class because after school, he went to work to help his single mom pay the rent. Another student confided in me that he was born with HIV/AIDS; this illness affected the rate at which he processed information. It took more time for him to read and understand the classwork. The message these two high school students were inadvertently telling me was that I was maintaining this inequitable structure of the U.S. educational system. If I was to be the teacher that these and all my students needed, I—not my students—would be the one that needed to change perspective, shift course, and become more attuned to the critical concerns my students lived with and endured while trying to pursue academic achievement.

2.5 Sunlight and Nutrient-Rich Soil for Every Plant: Teaching with Equity and Justice

> …education and freedom have been inextricably bound in black American life. (Givens, 2021, p. 13)

A thriving garden will have enough soil, nitrogen-fixing bacteria, sunlight, and water that each plant needs to the degree with which they need these things. Some plants prefer more shade than others, and some do better with composting materials. A knowledgeable gardener pays attention to each herb, adjusting the amounts of fertilizer, nutrients, and sunlight/shade. As an amateur gardener, I did my best to tend to my growing garden. However, I did not account for the sprawling mint plant that I was excited to have in my care. I loved fresh mint! I enjoyed being able to snip off a crisp leaf of this green, bright, aromatic herb. What I quickly learned about mint plants through Google searches and my in-the-moment experience, was that while a popular herb to grow easily at home, it is greedy. Mint plants sprawl across the surface of the ground, stretching it horizontally, inevitably crowding out the growth of other plants. Seemingly overnight, my mint plant was making its way near the cucumbers and basil, making its presence known. Similar to the emergent and canopy trees in a rainforest, blocking sunlight from the understory and forest floor plants below, my mint plant was taking whatever space it wanted to the detriment of the other members of my small agricultural community. Despite the fact that mint has anti-pesticide properties that could benefit its neighbors, it wasn't playing fair nor was it too concerned about sharing its resources that could help others. And I would be complicit in this inequity and injustice if I continued to let it sprawl where it pleased. In my discovery of what was best for all of the diverse plants I was taking care of, I had to move the mint to a part of my garden where its sprawl was least likely to inhibit others—its own pot.

As with most analogies, this mint example from my garden breaks down and cannot adequately represent all facets of the real context like teaching and learning. But it does point to the awakening I had about teaching Black and Brown learners; often their brilliance is overshadowed and diminished because they are not given what they need, but what the "norm" has determined is best. In reflecting upon my journey in science education, I can see my focus begin to shift more towards the how of teaching (pedagogy) than solely the what (content), towards viewing teaching as an act of social justice that is equitable to all learners, particularly those who had been marginalized because of their race and ethnicity (Abrams et al., 2014; Freire, 1970; Le et al., 2019; McCurdy et al., 2021; Morton et al., 2022).

Like many Black children who grew up in suburban communities, I was the only Black girl, and sometimes Black student, in most of my classes as I attended predominantly White schools, both public and private. I had very high expectations for my learning, education, and behavior. I strived for excellence, the best grades, knowing the answers, and studying to make sure I was prepared for class and any tests I would have. Learning and understanding were important to me.

I was also familiar with my family's role in Civil Rights from a young age. I grew up hearing my mother telling me and my brothers about her participation in the early 1960s student-driven sit-ins and Civil Rights Marches with Dr. Martin Luther King, Jr. while attending Fisk University in Tennessee, and marching again in St. Augustine, Florida joined by her mother-in-law. Her father, my grandfather, Rev. Thomas A. Wright, Sr., a Black Baptist pastor, played a significant role in establishing and leading the National Association for the Advancement of Colored People (NAACP) and other Civil Rights Movement events in St. Augustine, Florida. In 1965, my aunt was the first Black student to integrate and graduate from Gainesville High School in Florida. This rich family heritage became a large part of my foundational belief that people's skin color, socioeconomic status, or gender should not prevent or impede their equitable access to opportunities, resources, or rights in any way. I come from a long line of teachers and educators including grandparents, aunts, uncles, and especially my parents, all of whom practiced and knew the value of effective, persistent, and empathetic teaching to provide underrepresented students an opportunity to achieve more than prior generations. These equity-oriented educational pioneers were instrumental in shaping my view as a learner, a teacher, and as one who needs to make sure the voices of the unheard are brought forward to speak into the microphone that has largely been passed right by them. All of these stories and the legacy of my forebearers somehow seeped into me, into the air around me as I grew into adulthood.

However, I became more sensitive to the ways in which my privileged educational and justice-oriented heritage had shaped me as I began teaching. Herein lies some of the lessons I learned as a young teacher of ethnically diverse adolescent learners at Title 1 schools. I became aware that while many of my students had similar amounts of melanin, racial, and ethnic ancestry as I did, the family, community, and socioeconomic contexts often varied from my own. My parents provided me and my brothers with a level of access and opportunity that many of my students did not have. I had an awakening of sorts that I, a Black woman, could not always relate to my students' work ethic, or behavior, or vernacular. I placed the onus onto them. In other terms, I, a Black woman was viewing children who looked like me from a deficit lens. What one does not always learn in education courses is that inherent bias swirls around in educational systems and teacher preparation programs. The story of science and that of education in the United States has been steeped in a Euro-centric, hierarchical, classist environment (McCurdy et al., 2021). It can grip anyone of us and permeate through our teaching and learning without anyone taking notice.

I taught full-time for 13 years at Title 1 middle and high schools (in which the majority of students come from low-income households and are provided with free breakfast and lunch) in two school districts in the state of Florida. I did not begin teaching science with the expressed intention of being a role model for students of color. My goal was to help students better understand the importance and functions of the cell, the development of organisms and the processes that sustain them, and the interactions that occur between living things and the natural world. My goal was to create a love of science in my students that I always had. But goals can sometimes change. In fact, sometimes, goals must change in order for a greater purpose or mission to surface.

While teaching seventh-grade science, a Black female student who was moved from my classroom to another teacher's class requested that she be placed back into my class. I explained to her that her current teacher was good, and that she'd do well in his class. She told me, "Mrs. McCurdy, I've never had a Black female teacher before, and I really would like to stay in your classroom to have you as my teacher." Whether or not she was saying this so she could be with her friend in my class, I do not know. Maybe both reasons were true, but her words stay with me because regardless of what I am teaching or presenting to students, they see me, what I look like. For this student, having a teacher that looked like her mirrored a part of who she was and what she aspired to be in some way.

At that point, my Blackness was most relevant when I taught Black students. I did not let them off the hook for effort, academics, or behavior. There have been many heart-to-heart "chats" with a student or two after class or in the adjacent planning room to remind students of color who they were, what they looked like, and how the world viewed them. I knew that no other teacher could (a) speak to them from similar racial experiences, (b) call out the concealed excellence in them, or (c) frankly call attention to the reality of their world outside of the classroom. I do not know if I can ever *just* teach science anymore.

I was learning, as my mentor told me and as his mentor told him, that my presence, my Blackness, in a classroom speaks for me before I open my mouth. I was learning that I needed to leave the K-12 science classroom to prepare future teachers of science to teach with equity and relevance too.

2.6 Digging Deeper Roots: Leaving K-12 and Entering Science Education PhD Program

Black butterfly, sailed across the waters
Tell your sons and daughters
What the struggle brings
Black Butterfly, set the skies on fire
Rise up even higher
So the ageless winds of time can catch your wings
Let the current lift your heart and send it soaring
Write the timeless message clear across the sky
So that all of can read it and remember when we need it
That a dream conceived in truth can never die
Butterfly

'Cause now that you're free and the world has come to see
Just how proud and beautiful you are
—Performed by Deniece Williams (Weil & Mann, 1982)

One of the delights of having a garden or being in a garden area is seeing the butterflies flittering around the colorful flower petals of marigolds, opening, closing, and flapping their wings after landing on just the right spot. After being in the classroom for close to 14 years, I knew that it was time for me to venture outside of the cocoon. Like a butterfly, I felt as though I could not be contained within the K-12 classroom, not because I did not enjoy it, but because I did, because that space is so significant. The nourishment I was receiving as I pursued my Educational Specialist degree in K-8 Math and Science Teaching was opening me up to the work that needs to happen on the other side, in undergraduate and graduate teacher preparation. My questions, my insights, my wonderings, and my reach were getting bigger and more expansive for where I had been. I needed to find a new home. This impending metamorphosis is captured in Fig. 2.3, an excerpt of my teaching statement I wrote in 2017 as part of my application to the Science Education PhD program at a research-heavy university in southeast United States.

2.7 Diversity in the Garden of Science Education

When I started researching how to start a garden, one goal I had was to explore those herbs and vegetables that would help me with a health issue I was having, stemming from stress and mild anxiety. I began researching plants that can be planted together or near each other to make use of the small space I had laid out for the garden. I discovered that there are certain plants, herbs, and flowers that are companions to one another. They provide a diversity of nutrients to the soil, help keep pests away,

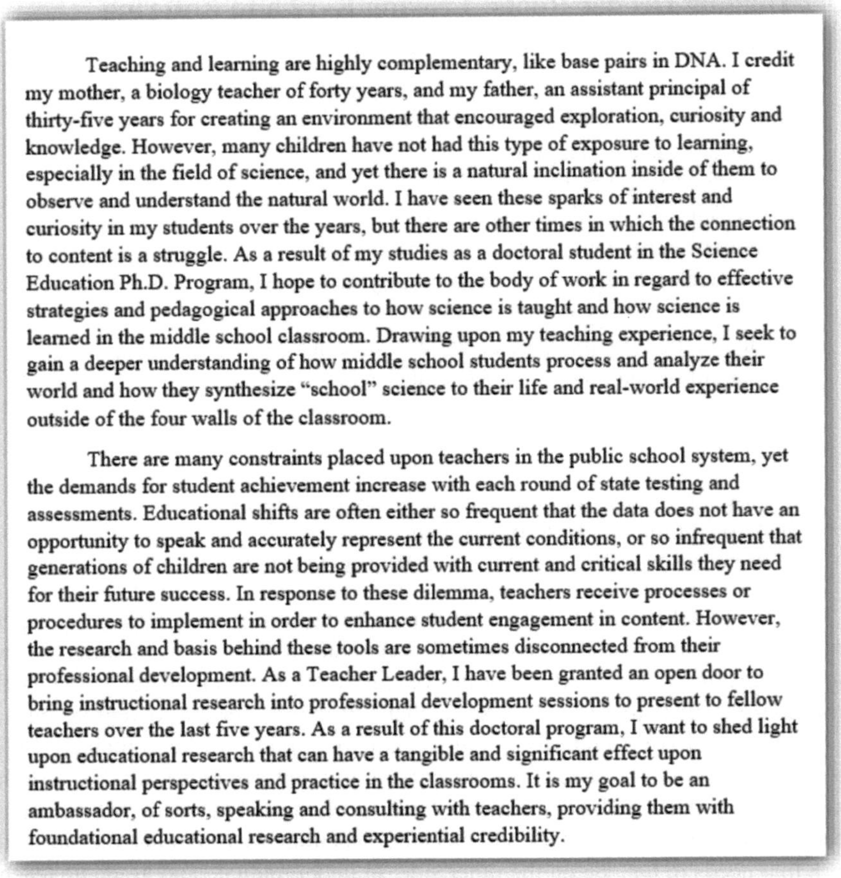

Teaching and learning are highly complementary, like base pairs in DNA. I credit my mother, a biology teacher of forty years, and my father, an assistant principal of thirty-five years for creating an environment that encouraged exploration, curiosity and knowledge. However, many children have not had this type of exposure to learning, especially in the field of science, and yet there is a natural inclination inside of them to observe and understand the natural world. I have seen these sparks of interest and curiosity in my students over the years, but there are other times in which the connection to content is a struggle. As a result of my studies as a doctoral student in the Science Education Ph.D. Program, I hope to contribute to the body of work in regard to effective strategies and pedagogical approaches to how science is taught and how science is learned in the middle school classroom. Drawing upon my teaching experience, I seek to gain a deeper understanding of how middle school students process and analyze their world and how they synthesize "school" science to their life and real-world experience outside of the four walls of the classroom.

There are many constraints placed upon teachers in the public school system, yet the demands for student achievement increase with each round of state testing and assessments. Educational shifts are often either so frequent that the data does not have an opportunity to speak and accurately represent the current conditions, or so infrequent that generations of children are not being provided with current and critical skills they need for their future success. In response to these dilemma, teachers receive processes or procedures to implement in order to enhance student engagement in content. However, the research and basis behind these tools are sometimes disconnected from their professional development. As a Teacher Leader, I have been granted an open door to bring instructional research into professional development sessions to present to fellow teachers over the last five years. As a result of this doctoral program, I want to shed light upon educational research that can have a tangible and significant effect upon instructional perspectives and practice in the classrooms. It is my goal to be an ambassador, of sorts, speaking and consulting with teachers, providing them with foundational educational research and experiential credibility.

Fig. 2.3 Doctoral goal statement document

attract pollinators for other crops, and support their neighboring plant companions as they grow.

In North America, Native or Indigenous people have a long history of growing the Three Sisters crops or corn, beans, and squash in the same space. In her book, *Braiding Sweetgrass* (2013), Robin Wall Kimmerer shares how each of these provided some vital aspect to development that the others needed. Corn's height and tall structure gives support for the bean once the bean begins its vine-like ascent toward the sun. The squash extends itself horizontally along the ground, using its leaves to cover the soil to maintain moisture and keeping caterpillars away with its prickly vines. As the roots of the bean descend further into the depths of the soil, nodules form on the roots and soon become home to nitrogen-fixing *Rhizobium* bacteria. These microscopic tenants engage in a mutualistic relationship, converting nitrogen in the air into usable ammonia-based natural fertilizer. This polyculture, though diverse in form, structure, and physicality, are less susceptible to pest infestations than monocultures. "The gifts of each are more fully expressed when they are nurtured together than alone" (Kimmerer, 2013, p. 140).

At the time of this writing in the United States, 2024, much is at stake for the world of science education, and for issues of equity and justice as it pertains to people of color and our history in this country. In telling this journey about my home in science education as a Black scholar, I thought it necessary to add how my thoughts and approach to science teaching and learning evolved from my early years of teaching to my pursuit of an assistant professor of science education. As I re-read these excerpts from the diversity statement I wrote for professor positions, it reminds me of why I am in this field, doing this hard but necessary work.

When I entered teaching, I was ignorant of the ways diversity played a role in my teaching and students' learning. As I reflect in Fig. 2.4, I began to view my background differently and therefore I was able to recognize the unique cultural assets my students had.

The fabric of science teaching within the United States and other westernized societies has been so inundated with one single narrative of science, that students of all races and ethnicities sitting under this type of instruction can live out their entire lives not knowing about the innumerable contributions to the sciences of people of color globally (Iaccarino, 2003; McCurdy et al., 2021; Ogawa, 1995). The idea of homogenizing all of our diversities can become a dangerous norm that can distort cultural identity. In this section of my diversity statement, found in Fig. 2.5, I attempt to wrestle with this as it pertains to my international and national learning contexts.

A few months before writing my diversity statement, I facilitated a virtual colloquium professional development with the purpose of helping STEM faculty at a local four-year college develop relevant inclusive learning environments. I developed the graphic in Fig. 2.6 to visually convey what my words encapsulate.

This graphic, what I am calling the Multicultural Wheel of Science and Ways of Knowing, communicates that much of Westernized science that is taught hides its true hues that are more indicative of Black and Brown people across time. Even though there are countries and people groups missing in this image, it signifies that

> My positionality as a Black woman growing up in predominantly White suburban community in South Florida and attending predominantly White institutions from K-16 critically shape how I live as a teacher, student, colleague, researcher, and a global citizen. I do not assume that the voice of the "majority" represents my own, as their experiences often do not include the history of marginalization, underrepresentation, and misidentification that women and other traditionally marginalized individuals internationally have had and continue to encounter. The purpose of my teaching practice and research, therefore, is to emphasize education's ability to empower all learners and show teachers—and future teachers—the necessity of and value in including all voices and perspectives in the classroom.

Fig. 2.4 My recognition of diverse voices in my diversity statement

There is an assumption that diverse students within one race, culture, or group are a monolith, however it is this belief that continues to permeate classrooms, distancing marginalized students from their teachers and the content being presented. The time I spent living in Mozambique, Africa teaching English showed me that even though individuals may share similar outward appearances, their societal upbringing can greatly shape the lens through which they live and understand the world. In *Pedagogy of the Oppressed*, Paulo Freire affirms the need for the classroom to encourage dialogue between teacher and students and among students to develop critical thought and problem-solving actions that relate to students' real lives and experiences.

Fig. 2.5 Description of diverse societal perspectives (from diversity statement)

Fig. 2.6 Multicultural wheel of science and ways of knowing

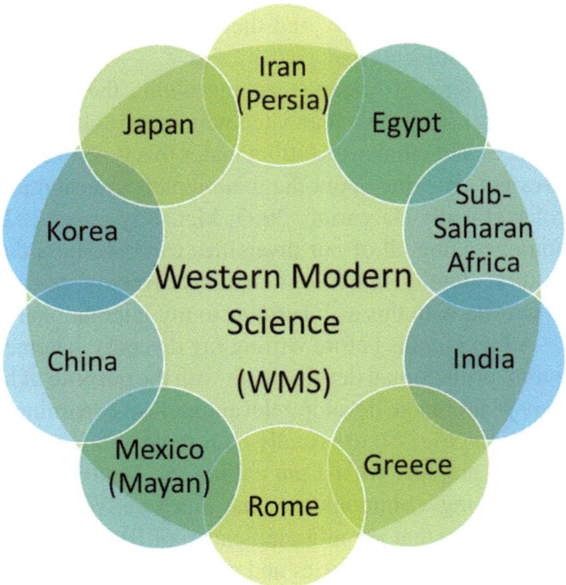

idea that science/STEM teachers should be "taught how to leverage the rich history of the myriad voices that have shaped and continue to shape science content and ways of thinking so that students see a reflection of themselves in this culture of science" (McCurdy et al., 2021, p. 13).

During this professional development, I challenged the instructors to take note of their blind spots within their subjects and their teaching methods. In the journey through my garden, I had to take note of what the plants were communicating to me. If they were too dry, or if I watered them too much, I needed to be attentive and respond accordingly. Similarly, my pedagogical approaches of a more seasoned K-12 teacher shifted from my early teaching days. I diversified my approaches to continually worked to shift the narrative from historical Westernized ways of teaching and learning science to that which centers the needs and perspectives of the learners who come from various races, ethnicities, genders, exceptionalities, and socioeconomic statuses. Examples of these instructional shifts included collaborating with other teachers to expose middle school girls to computer science at a local college, planning 8th grade STEM-focused field trips for students to gain tangible ways of using STEM outside of the physical classroom, and implementing projects for students to develop their own problem to solve by incorporating their cultural experiences. The classroom learning environment became less about me as a teacher and the knowledge I had to give and more about creating a student-centered space where diverse authentic student thinking could thrive.

Fig. 2.7 Shift to student-centered teaching (from diversity statement)

lesson planning design with my learners in mind. The excerpt in Fig. 2.7 peeks into this transformation I experienced.

Like many full-time doctoral students, every semester of my program, I worked as a graduate research assistant and/or a graduate teaching assistant. This was an open field which I could till the soil; I incorporated and modeled culturally relevant and equity-oriented teaching practices. I started weaving in stories of women and people of color in the science activities, challenging these future teachers to broaden their view of who does science and what scientists look like.

Classrooms around the world are becoming more culturally and ethnically diverse. Therefore, effective equitable teaching and learning needs to include: (a) developing a comprehensive understanding of the research behind the formation of the science identities of underserved and underrepresented student populations in primary and secondary classrooms; (b) providing meaningful research and practice-driven professional development for in-service teachers of science; and (c) mentoring and instructing science teacher candidates towards the development of culturally relevant science instruction (Ladson-Billings, 2006) that promotes the active learning of the nature and practices of science in relevant ways. I grew up learning how to relate to, interact with, learn from, and collaborate with people from all different racial and cultural backgrounds. Diversity, inclusivity, and equity are not simply aspects of my professional roles; these ideals are a part of who I am and how I move about in the world. They take intention and commitment to all science learners, because "we ultimately have a lifelong impact on who [learners] become and the kind of society in which we all will ultimately live" (Ladson-Billings, 2006, p. 40).

2.8 Pruning, Uprooting and Replanting: From PhD Student to Assistant Professor

> A farmer went out to sow his seed. As he was scattering the seed, some fell along the path; it was trampled on, and the birds ate it up. Some fell on rocky ground, and when it came up, the plants withered because they had no moisture. Other seed fell among thorns, which grew up with it and choked the plants. Still other seed fell on good soil. It came up and yielded a crop, a hundred times more than was sown. (The Parable of the Sower, Luke 5:5-8, NIV Translation of the Bible)

The processes of preparing the ground, sowing seed, watering the soil, monitoring growth, or detecting disease or the presence of pests are expectations of someone who farms or gardens. There are times when the harsh climate conditions challenge these processes, and the farmer is forced to dig up certain plants or crops to located them, to better care for the soil, and the like. There are other times when the farmer stands back with her hands on her hips and smiles in hope and pride.

The Parable of the Sower offers a description of the purpose of the farmer: to sow. This is central to their role. Given this job description, the farmer is aware that challenges with the seeds may be detrimental to the growing plant or crop. She needs to be ready to prune, uproot, and replant should the circumstance arise. Transitioning from PhD student to Assistant Professor was undoubtedly rewarding. I was pruned, and I knew I had to move to one day take a new position to continue my purpose: to, teach, research, serve, and mentor. This last section of the chapter shares touchpoints that I can look to when I need to remember why I do the work of science education and pass it forward whenever I can.

2.9 Mimicry: The Temptation to Compare or the Ability to Transform in Academic Pursuit

In the natural world, certain plants, insects, and animals present themselves as different organisms in myriad ways to sustain and advance their survival. Some organisms appear to mimic predators as a protective measure to prevent becoming prey themselves. Whatever the situation or scenario, these creatures seem to have an advantage in their environment because they have fooled others into believing they, the mimics, are something that they are not. It is possible that in every garden there is some type of mimicry taking place, but not all mimicry is advantageous, especially as it pertains to my journey in science education.

Having started my doctoral studies in my early 40s, there were times when I felt that I could never catch up. I came in very focused. I was fully committed. I resigned from my middle school teaching position, and became a PhD student full-time. I knew what I wanted to research for my dissertation work. I was on a mission. I had a very supportive husband and young children. Many of my professors and mentors were the best I could have had. The work and long nights were tough at times, but I was persistent and purposeful. Until comparison crept in.

When projects and tasks were moving quickly, it was hard to detect what was (and sometimes still is) lurking around the corner. In the still moments, I would start to look around to what my peers looked like, what they were doing in their research, what opportunities they had that I did not have. I had a family and responsibilities that came with this. The allotment of my time, therefore, was usually divided for me. I was not as young. My schedule was not as available. I was not as flexible to spread myself as much as they did.

I am sure the idea of me as a Black woman played a part in this as well. I had heard all of the facts and numbers about how many women had the PhDs. There were even fewer Black women with PhDs. It was very easy to think that I was simply mimicking the process the whole time, especially amidst the late night or early morning writing sessions. Persistence would be continuing to take steps in faith, even though the odds did not appear to be in my favor.

In Zora Neale Hurston's novel Their Eyes Were Watching God, the character Janie says, "There are years that ask questions and years that answer" (Hurston, 1937, p. 38). The temptation to compare sometimes led my mind to entertain a bevy of "What if" questions. *What if I can't find a tenure-track position? What if I never get a manuscript published? What if...* These questions can be stifling. The temptation to compare could have led me to engage in mimicry to cover seeming inadequacies I felt at times.

What illuminated and encouraged me during those moments were my faith in God, my family support, and an openness to develop. Throughout my doctoral years, I continued to attend my home church, participate in small Bible study groups, pray, and trust God. I remembered that this calling to academia was real and had been confirmed in many ways.

My husband and two elementary aged children were instrumental in reminding me of my worth in this space. While steadily working on my dissertation during the pandemic, my nine-year old son was upset that I had not seen his drawing yet. At this point, my husband came in to save the day. He said, "Mommy is in super-school, and at the end of super-school, Mommy would have to write a big book, but without pictures." I turned to my son whose face seemed to brighten a bit, and told him "It's true." However, I took it a step farther. I decided to brand my superhero status shown in Fig. 2.8.

Lastly, it was critical for me to be authentic and real with myself. Like every other academic pursuit, doctoral students should be reflecting, learning, and developing. We do not enter academia with all of the knowledge we need; at times, we need to be exposed to other perspectives to refine and critique our own. Figure 2.9 shows an image from a note in my phone app of some musings I had after a year in the PhD program that sheds light on this point.

While all mimicry is not negative in nature, it was important for me to know who I was, who I was not, what I knew for sure, and what I was still figuring out in order to be myself as I took the next step. Finding a job and getting to work as an assistant professor!

Fig. 2.8 My Bitmoji in superschool (PhD program)

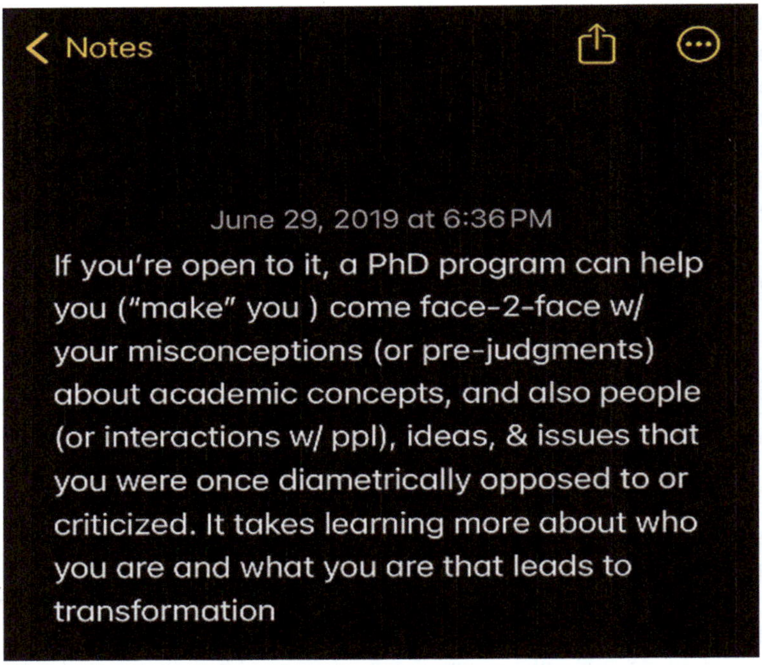

Fig. 2.9 Image of phone note describing PhD Program

2.10 Thorns, Thistles, Invasive Plants, and Squirrels

Sometimes in the garden there are thorns. Sometimes there are intruders. Sometimes there are other plants that pop up unknowingly from pollen that's been dispersed from other places and lands on the flowers and the plants that I want to grow. In the center of the garden that I was tending to, there was a mango tree. And for the first couple years, it did not fruit very much. But then, after it was pruned, it began to flourish and drop mangoes during the season. And as someone from South Florida, mangoes are a part of my childhood summer enjoyment, the sweetness that they have and the many ways they can be used in culinary dishes. When I think of mangoes, I think of my mother gathering them from family members' homes nearby or friends and neighbors who had mango trees and even our own mango tree in the backyard of our house. I think of her, me, and my brothers peeling mangoes and slicing them into bowls and dishes and laying them out for us to enjoy. On a hot summer day, we may put mangoes in a bowl of scoops of cold vanilla ice cream. Having a mango tree in my home as an adult, as a wife and mother was a way to share and relive some of those childhood memories with my own children. Cutting mangoes and having my own children eat them. One summer, we had so many mangoes in our backyard that we shared them with our neighbor and relatives. However, one of those summers we also had critters who were taking the mangoes. Squirrels. They were feisty and would get to the mangoes right at the moment that they were ripe and ready for us to pick. And within these mangoes were bite and claw marks on partially eating mangoes left all over the backyard. While squirrels are not necessarily enemies of mine. They were taking things that were cherished items to me and my family.

This somewhat comical scene of squirrels and mangoes does remind me that not everyone views my work, scholarship, and my teaching with the same acknowledgement, value and understanding as I do. There have been moments through my journey as a doctoral student and as an early career faculty member that I have had to remind myself that the work of equity, diversity and authentic culturally responsive science teaching and learning is not welcomed by some and maybe many. There was an incident at an event in which I was leading a workshop for more than 100 faculty members; I was describing how they can help make their subjects and course more inclusive and relevant to all of their learners. Most of the faculty in the workshop seemed to be on board. They were participating and asking good questions in their small groups. Then one of the professors stood up. He was not necessarily asking a question but was trying to refute what I was intending and hoping to convey to everyone. I do not recall his exact words, but he clearly was pushing back against being receptive to incorporating examples, insight, and course materials from people of color and Indigenous ways of knowing into his curriculum. After the workshop, and after the keynote speaker spoke to this professor on my behalf, unbeknownst to me, the professor came to offer a "I did not mean any disrespect, but…" comment to save face.

Yet, he was taking these mangoes from the tree and not even eating them. He was tearing them apart publicly, and throwing away any part of the seed so that it would not take root anywhere. It was a good reminder for me that there are aspects to my work that everyone will not appreciate. There are aspects of my work that are more sacred or special than others see, and I do not have to allow everyone into that space of my work. I am not necessarily welcomed into every space of others' work. Even within the classroom, there are students who are skeptical of my accomplishments and have said such in their anonymous university-level course evaluations. It is good to be self-reflective, receiving critical feedback that is intended for growth. But unsolicited and unhelpful criticism can obstruct the purpose. To this end, I am learning that it is necessary for me to be surrounded by meaningful others who understand this work, who are doing this work, or who support and enrich me and the journey of my garden.

2.11 Mast Fruiting: Staying Connected and Accepting Opportunities, and Mentoring

> Some studies of mast fruiting suggest that the mechanism for synchrony comes underground. The roots of trees in a forest are often interconnected by networks of mycorrhizae and fungal strands…They weave a web of reciprocity, of giving and taking…the trees all act as one…Through unity, they have survival. (Kimmerer, 2022, p. 37)

When I first began considering pursuing a PhD, Dr. Baobab (pseudonym), the head of the science education program, provided me with sage insight, as is symbolized by his tree pseudonym. He was wise, supportive, and many found refuge and encouragement when they spoke with him. One of the things he told me, was the PhD is more than classes or coursework, but it is a program and it will change my life and that of my family. This was very accurate.

Dr. Baobab also told me that I need to find *my* people. He introduced me to other PhD students, to inspirational faculty of color, and to communities, like the Holmes Scholar Program, that would feed the soul of my culture and my work. He knew, being a Black man, that staying connected with deep roots in good soil would be part of me surviving and thriving in academia. From those connections with Holmes Scholar, I met brilliant doctoral students of color who do this work at their institutions. From this community, I met two women who reached out to connect with me through Zoon and write together. Over the years, we have met, shared updates on our personal and professional lives, and presented the collaborative work we developed at a STEMNoire Conference, an organization for Black women in STEM.

I have had opportunities to meet other strong, admirable, and influential individuals, like Dr. Aspen and Dr. Sycamore (pseudonyms), in the larger science education community through the mentorship Dr. Baobab gave me. He served (and still serves) as a connector, communicating to these other strong trees, "I think you need to get to know Regina." Then seemingly overnight, opportunities opened up before me. Opportunities to write a response article in widely read journal. The chance to review an article in another top science education journal. I was never short of options and doors being open, largely due to Dr. Baobab seeing something in me that he knew needed to be cultivated and shared. But the other side of these opportunities was a decision that only I could make. I had to say "Yes". I had to decide whether I was going to stay safe and mimic others, or to be transformed and blossom into some more multilayered tree with many more roots and branches than I had before. I chose the latter, and I know my garden is the better because of it.

The plants, trees, soil, and critters that make a home in the same space reiterate the idea that "All flourishing is mutual" (Kimmerer, 2013, p. 15). I was blessed to not have to navigate the ins and outs of academia as a Black woman, wife, and mother alone. And so, as I replanted myself in another geographic location, I have taken the lessons of the garden and sow seeds as I am able to nourish the next generation of science education scholars. As I have received wonderful mentorship during my graduate years, I look for opportunities to provide whatever support and guidance I can pass along to make the next generation of science education scholars confident in the gifts they have to share through their thinking, work, and service. In so doing, I carry on the legacy of my ancestors, family, teaching and learning community, thankful that I had so much given to me, so I can give it away.

References

Abrams, E., Yore, L. D., Bang, M., Brayboy, B. M. J., Castagno, A., Kidman, J., Lee, H., Villanueva, M. G., Wang, M. H., Webb, P., & Yen, C. (2014). Culturally relevant schooling in science for indigenous learners worldwide: Stressing the all in science literacy for all. In N. G. Lederman & S. K. Abell (Eds.), *Handbook of research on science education* (Vol. 2, pp. 671–696). Routledge.

Freire, P. (1970). *Pedagogy of the oppressed* (M. B. Ramos, Trans., p. 2007). Continuum.

Givens, J. R. (2021). *Fugitive pedagogy: Carter G. Woodson and the art of black teaching*. First Harvard University Press.

Hurston, Z. (1937). *Their eyes were watching God*. University of Illinois Press.

Iaccarino, M. (2003). Science and culture. Western science could learn a thing or two from the way science is done in other cultures. *EMBO Reports, 4*(3), 220–223. https://doi.org/10.1038/sj.embor.embor781

Kimmerer, R. W. (2013). *Braiding sweetgrass: Indigenous wisdom, scientific knowledge and the teachings of plants*. Milkweed Editions.

Kimmerer, R. W., Gray Smith, M., & Neidhardt, N. (2022). *Braiding sweetgrass for young adults: Indigenous wisdom, scientific knowledge, and the teachings of plants*. Zest Books, an imprint of Lerner Publishing Group, Inc.

Ladson-Billings, G. (2006). "Yes, but how do we do it?" Practicing culturally relevant pedagogy. In J. G. Landsman & C. W. Lewis (Eds.), *White teachers diverse classrooms: Creating inclusive schools, building on students' diversity, and providing true educational equity* (pp. 33–46). Stylus.

Le, P. T., Doughty, L., Thompson, A. N., & Hartley, L. M. (2019). Investigating undergraduate biology students' science identity production. *Life Sciences Education, 18*(50), 1–16.

Longstreet, W., & Shane, H. (1993). *Curriculum for a new millennium*. Allyn and Bacon.

McCurdy, R., Cruz-Deiter, L., & Butler, M. (2021). Preparing teachers of science for the multicultural classroom through a global lens. In M. M. Atwater (Ed.), *International handbook of research on multicultural science education* (pp. 1073–1104). Springer.

Morton, T. R., Miles, M. L., Roby, R. S., & Ortiz, N. A. (2022). "All we Wanna do is be free": Advocating for Black liberation in and through K-12 science education. *Journal of Science Teacher Education, 33*(2), 131–153. https://doi.org/10.1080/1046560X.2021.2008096

Ogawa, M. (1995). Science education in a multi-science perspective. *Science Education, 79*(5), 583–593.

Weil, C., & Mann, B. (1982). Black Butterfly [Recorded by Williams, D.]. On *Let's hear it for the boy* [Record]. Columbia Records.

Part II
Found a Home

Chapter 3
Drowning Out the Background Noise

Brenda Brand

3.1 Introduction

When making the move from being a science curriculum administrator to becoming a science educator in academia, I contemplated this shift in my identity and how my sense of purpose would be redefined. I decided very early after assuming this role that my work would not become entangled in a battle of significance. When assuming this new position, I was immediately confronted with the demands of academia. I realized that if I was going to succeed in this environment, I needed to process and declare my own identity as a scholar. I drew from my academic training and family upbringing to condition myself for my new role. My first dilemma was to define the word "success" for ME.

In academia, much pressure is precipitated by the struggle to identify with the images of significance, productivity, and success. This pressure can influence your goals, work habits, self-esteem, confidence, and overall work experience. That is what I refer to as background noise. It is echoed in every meeting, every conversation in passing, and even during quiet times, whether it is at home, in the office, during leisure activities, everywhere.

Added to the expectation of demonstrating a consistent record of scholarship, there is an additional layer of pressure to navigate the sociocultural implications. These are societally induced pressures that insinuate a need to counter biased norms about people who look like you (the voices only you can hear), all while operating in a system that is a breeding ground for comparisons. You must declare your worth against a narrative of socially prescribed descriptions of what constitutes quality research. You have the pressure associated with doing the work, and the mental and

B. Brand (✉)
Virginia Tech, Blacksburg, VA, USA
e-mail: bbrand@vt.edu

M. M. Atwater and B. A. Garvin (eds.), *Creating and Maintaining a Home in Science Education Research*, Contemporary Trends and Issues in Science Education 65,
https://doi.org/10.1007/978-3-031-99382-4_3

emotional taxation to convince everyone of the meaningfulness and significance of your research agenda.

Add to that, you are a qualitative researcher, and the recommendations that were coming through the department did not seem to consider all of the work that qualitative methodology entails, the collection of the data, the transcribing of the data, data analysis and the review process. As a junior faculty, I was not prepared for the task of juggling the qualitative process against the messages of what it meant to be a scholar, realizing the meanings that I wanted to convey through research could be considered a counter productive use of my time. After all, it was my decision to be a qualitative researcher. I was told by one colleague that it did not matter why something occurred, it was enough just to know that it occurred.

My value for qualitative research began with the first qualitative research study that I conducted, which I will refer to as the pilot research. I was asked by my advisor to conduct a research study for an ongoing funded project investigating factors influencing learning in science and mathematics for students from underrepresented groups. Five students who were participants in a project that identifies future educators while in high school, the "Future Educators Program" (FEP) were interviewed (Brand et al., 2006). One of the students was an undergraduate already enrolled at the university. The others were seniors in high school. I asked open-ended questions to obtain the students' perspectives on their learning in STEM. My sense of purpose for my scholarship emerged from the findings of this study. The students' discussions primarily focused on stereotypes and how stereotypes associated with their identities in science and mathematics imposed limits on expected outcomes related to their achievement. The findings highlighted the value of relationships with teachers to helping them navigate social constraints and move beyond them.

One day, in a project meeting, a faculty member in mathematics from the university pulled me aside and asked if he increased technological applications in his mathematics course at the university, would he be placing students from underrepresented groups at a disadvantage. While I was stunned by the question, I assured him that the students would not be disadvantaged by curricular modifications that included technology. His question corroborated the impact of stereotypical messaging discussed by the students in the pilot research, signifying the importance of qualitative research in representing the voices of the students in exposing the impact of marginalization on their education. I realized that qualitative research, as well as examining statistical relationships, should center the voices and experiences of the participants to add meaning and determine effectiveness. Data in this pilot research brought into focus the belief systems and the sociocultural nuances that influence underrepresentation. The data from this research highlighted students' cognizance of racialized and gendered identities that are front and center when focusing on marginalized students' achievement in science and mathematics.

3.2 Meanings from the Pilot Research

I was inspired by the views shared by the students from the FEP program. During my schooling, the majority of my teachers were African American, and the predominant messages focused on being proud of our identity and working hard to prove our competence. While these messages were indeed resistance to circumstances voiced by the students, the emphasis from our teachers focused on defying odds and beating the system. We were aware that there were divides, but the atmosphere was more focused on the history of African American oppression, and achieving justice through overcoming challenges and obtaining an education. We were taught that everything we did was a reflection on our community, and that in society, little was expected from us, so we needed to go above and beyond in our accomplishments to gain any level of credibility. Fast forward, I attended a predominantly White university as an undergraduate, and I was guided by the urging to demonstrate my competence and reflect positively on my community. Conducting this study in the early stages of my doctoral studies proved to be a meaningful experience. The students' views on how their educational experiences were impacted by stereotypes and deficit narratives caused me to reflect on my own life's experiences and how I was conditioned to oppose social narratives.

Bourdieu (1997) introduces capital to explain the presence of resources that impact the ease at which individuals navigate the social world. Bourdieu (1997) asserts that capital influences individuals' positioning in society, and it is a myth to claim that capital does not contribute to advancement. Bourdieu (1997) introduced three forms of capital: economic, social, and cultural capital in three states, embodied, objectified, and institutionalized. Bourdieu (1997) explains economic capital as the abundance of finances and material wealth an individual possesses or has access to. Social capital is explained as a community or network through relationships that grant access to certain resources and opportunities. Cultural capital exists in three forms, embodied, objectified and institutional. Embodied cultural capital is the habitus that you receive from your upbringing, the language that is spoken, the collective ways of being. Objectified capital is ownership of tangible possessions that have value such as artwork, jewelry, etc. Institutionalized capital is status obtained from a certification or a degree that is granted by an institution. Bourdieu (1997) asserts that possession of any form of capital constitutes a leg up when aiming to assimilate and achieve in society. I reflected on the stories of the students and what would be considered their capital. I also contemplated my capital and how it contributed to my success.

Throughout history, differences in the United States were stratified through harboring economic and political power, resulting in the marginalization of groups and their communities. The history of African Americans is replete with stories of individuals who maximized the minimal in an environment that was restrictive and oppressive. Declared norms marginalized groups leaving them with the challenge of sustaining their lives in an environment that devalued them and their cultures. The circumstances resulting from these disparities are reflected in the accounts shared by the students in the PEP program.

Jeff (pseudonym), an African American male student worked in the evenings while attending high school. Even though he maintained an A average, he felt that his teachers would always judge him by the stereotypes associated with his gender and ethnicity. Leslie (pseudonym), believed that there were certain career pathways hinted for African American students, primarily fields in the entertainment industry, and that this was by design to prevent them from realizing their academic potential. Keith (pseudonym), concurred with Leslie, asserting that the media cajoles pipe dreams for African American youth to choose career paths that are high paying (professional sports and music) with low probability for being selected, further disenfranchising them in terms of financial stability. Greg (pseudonym), believed that African American students would avoid enrolling in certain STEM courses due to highly publicized messaging about their difficulty, and deficit narratives around African Americans' capabilities in STEM courses and career fields.

Bourdieu (1997) captured the fortuity of the advantaged and the plight of the disadvantaged through an interpretation of factors that advance success. Bourdieu's (1997) theory of capital yields insight for improving the achievement potential and livelihood of those who have been disadvantaged. Leslie's discussion highlights the capital in her background that influenced her achievement.

> I want it to be known that this guy sitting beside me is not the only person who is in the class. In math classes, I think that the teachers veer towards me because I demand the respect that I want in class, and they know that I have the backup which means my mom and dad. A lot of people don't, well a lot of minority students don't.

The support that Leslie received from her parents countered disparaging messages regarding her potential. Her parents were professionals which supported her in developing the habitus of schooling contributing to her academic success. She was empowered by her capital and recognized how individuals without capital were disadvantaged. Therein lies the significance of the theory and research. Recognizing the significance of capital influenced my interpretation of student outcomes and the messages that should be conveyed from this research. While at the time that this study was conducted, I had not been exposed to Bourdieu's (1997) theory of capital, looking back, I not only realized the significance of the findings to the field, but also to teacher education. Practically, I reflected more deeply on the role that teachers play in the lives of students from minoritized groups, and the purpose of role models and mentors. I integrated insights from the research into the science education program in an effort to facilitate the development of sociocultural consciousness in pre-service teachers that would encourage advocacy for students. I wanted to support them in realizing the need to become the capital that some students need to get them over the hump. These support systems translate into the guidance and support for historically marginalized students while painting a path for them to follow in order to achieve.

Representing the students' voices sheds light on the flaws of interpreting statistics on underrepresentation in STEM and the STEM achievement gaps solely based on numbers. The numbers may tell you how many, but they don't explain how the numbers came to be. Research findings undergirded by theory provide a context for interpreting factors that influence the percentages which translates into an awareness of social disparities and their root causes. This is key to interpreting and addressing under performance in historically marginalized students' academic achievement, particularly in STEM courses. Chubbuck (2010) describes three levels of analysis that teachers could engage in when making decisions on the appropriate strategy for addressing students' academic challenges: rudimentary, individual and structural. At the rudimentary level, the teacher develops a plan for remediating the content or skill for all students, which is a simplistic approach to interpreting the problem. It ignores external issues that may be contributing factors, and could result in blaming the student for not putting in enough effort to achieve. At the individual orientation level, the teacher focuses on identifying the student's challenges and develop strategies for addressing them. However, this level of processing could also result in blame being placed on the individual and/or their support systems. As a result, the teacher may decide that the extenuating circumstance is out of the teacher or school's capacity to solve and dismiss responsibility. Conversely, individual orientation processing recognizes that inequities exist and believe the problems can be fixed through individual support; however, there is limited consciousness around the larger systemic inequities. The third level of analysis is structural orientation. Reflections at the structural orientation level are also an individualistic approach to assessing the needs of the student; however, it includes a focus on identifying the underlying social infrastructure contributing to unequal outcomes for students. The teacher's response is informed by an awareness of the larger structural inequities that challenge students' achievement. For example, the teacher may recognize disparities in resources that place students at a disadvantage and make effort to provide the resources or reposition the student for access to the resources. Structural orientation analysis focuses on society, and not the students' home or family and acknowledges the far-reaching consequences of the disparity. Giroux (1992) states while it is important for teachers to listen empathetically to students of color, they need to understand how institutions of society influenced massive unemployment among African Americans, segregated schools, racist aggression, and substandard housing.

As a teacher educator I have integrated a social justice mission in the learning experiences for pre-service teachers. To foster sociocultural consciousness, I have been intentional about making preservice teachers aware of social practices that systematized unequal access to capital to obstruct paths to success for certain groups. For example, redlining or discriminatory lending practices marked certain communities as poor financial risks. Communities that were redlined consisted primarily of African Americans. They were discriminated against by mortgage lenders in an effort to sustain segregated communities. Redlining practices were bolstered by a notion that the presence of African Americans living within a community lowered the property value, insinuating a mental image that would bias how African Americans would be viewed in the larger society. To further disadvantage these communities, ensuing policies ensured that the taxes from these communities serve as the main source of funding for their schools. Consequently, schools in low socioeconomic communities depend on their tax base and as a result, are under-funded and under resourced. This agenda was designed to reinforce the marginalization of African Americans. It contributed to the deficit narratives that the students in the pilot research guarded against daily, which originated from the disparaging infrastructure as discussed by Giroux (1992).

Thus, advocacy for historically marginalized students amounts to creating a learning environment that shields them from deficit messaging associated with being an African American in this society. Teachers need to understand that stereotypes and lack of role models suggest to the students that they do not belong or have the capacity to achieve in STEM. They also need to understand that the measures that they take in the classroom to support underserved students' education is an effort to correct society's ill in creating an additional layer for historically marginalized students to plow through when attempting to learn in school. They are aiding students in championing invisible barriers that prevent equitable access and opportunities in STEM and providing learning experiences that support students in declaring themselves fit for STEM. Hussain and Jones (2021) assert that students feeling a sense of belonging and psychologically connected to their institution, is important to their well-being and achievement. Realizing a sense of well-being is particularly challenging for students when racial differences are unsupported at their institution (Foxx, 2021). Teachers who recognize their role as social justice educators are perceived by students to be in partnership with them against society. Jeff said it this way.

> It's hard for me to be a Black male and having to guard yourself. I'm in school now. I'm struggling hard to make this grade. It's just some teachers who don't believe that you are trying just because of what other people do.

Jeff's statement about what teachers believe highlights unspoken deliberations occurring between the teacher and the student as influenced by society's messaging. He believed that his teacher, in spite of his efforts to distance himself from the stereotypes, ascribed to negative messaging about his potential. Obviously, it was important to Jeff that his teachers viewed him outside of the stereotypes.

Conclusively, the importance of students' relationships with teachers was the major finding from the pilot research study. The students indicated that their academic performance depended on how they perceived their teacher's interactions with them. Tasha, one of the students at the university highlighted the barriers that prevents students of color from approaching their instructors, and how detrimental it is to achieving success.

> Like some of the teachers, there are certain ones where I've never been to their office. I never talk to them on the phone. I've never approached them after class. And those seem to be the classes that I'm having the most problems in, and they're not the hardest classes I have, which makes it so bad. Now, the classes that I have a lot of work in, that I have a lot of problems in, there's one professor, I am in her office at least twice a week. I'm always in there, and she's real easy to talk to.

Keith, like Tasha (pseudonym), also revealed the impact of barriers existing between teachers and students of color, explaining how the barriers inhibit relationships with the teacher, and how important positive student/teacher relationships are to achievement.

> Because if you get along with your teacher, I mean I had teachers I didn't like, and I didn't get good grades in the class. If you like your teacher, you are going to basically not be afraid to

> ask the teacher questions, and if you have a good relationship with the teacher, you will like to try to at least keep your grades up, so that you won't disappoint the teacher, or whatever. That's how I feel about it. The teachers I didn't like, I really didn't get good grades.

Significant are the meanings conveyed through these findings to the science classroom, and to contextualizing constructs such as rudimentary, individual, and structural orientation approaches for addressing students' challenges. The findings inform strategies that align with students' needs beyond just developing plans for remediating skills and knowledge. A teacher who uses a structural orientation approach for mitigating challenges with marginalized students in the classroom are aware of how racial underpinnings could negatively impact relationships between them and their students. They also recognize that racial tensions could result in students not trusting them. The teacher realizes that negative messaging about their identities ill position them academically, and the potential impact that it could have on their futures. A structural orientation approach looks beyond the surface of what could be perceived as resistance and understands that their off-putting interactions from students could be in response to nonverbal tension and the barrier that students hide behind to shield themselves from deficit views about their potential to succeed. This pilot study research was impactful because of all that I learned both personally and professionally from the findings which were revisited throughout my doctoral program. The meanings resurfaced over the years, and informed other projects and work, and while I have conducted other projects, this project laid the foundation and helped me to discover a sense of purpose as a science educator.

I consider myself to be a sociocultural researcher. Influencing sociocultural consciousness is my primary goal and is necessary for intervening in historically marginalized communities. On a larger scale, this consciousness translates into an awareness of society's reproductive power in procuring an unequal society which positions my work in the grander scheme. According to Apple and Weis (1986) schools play a significant role in the economic and cultural reproduction of the social strata, which is accomplished through messages conveyed by a hidden curriculum that legitimizes the norms and values of mainstream society. The hidden curriculum reinforces these norms and are considered the standard. Individuals whose experiences differ from the norms and values of the mainstream are disenfranchised and some respond by disengaging altogether, reaping the economic and social consequences of those who are ill prepared for achieving success in life. Apple and Weis (1986) also introduced the process of accumulation, which is the internal sorting of students by talent to secure an underclass. Students who exceed academic standards are positioned to excel and achieve a higher quality of life. This is compounded by the school funding policies that are designed to reinforce poverty in low socioeconomic communities which Jean Anyon (1980) explained as society's sorting and selecting process based upon the community in which a school is located. According to Anyon (1980), schools that are in affluent communities are preparing their students to occupy what would be considered more lucrative professional positions in society. On the other hand, students who attend schools in lower socioeconomic communities are relegated to either follow a vocational track or secure a working-class job.

Anyon (1981) asserted the power of social media's airwaves in communicating the messages of the hidden curriculum stating that it is not surprising that students internalize messages related to social class and roles more deeply than what is being taught in the formal curriculum.

Kohl (1994, 1995) introduced the state of "not learning," which is a student's response to a system that they think does not care about them or see them as unworthy of investment. Giroux (1981) revealed the irony in that decision, referring to it as resistance theory. Giroux (1981) pointed out that students who withdraw from their educational development shut themselves off from their ability to engage in social and political agency. They have limited possibilities for challenging the dominant ideology that is reinforced in schools, which is necessary for radical transformation. Freire and Macedo (1987) concurred. They spoke about the emancipatory nature of literacy, that by adopting the dominant syntax in society, students can meaningfully critique the world and their place in it.

In conclusion, while the African American students in the pilot research had obstacles to overcome, they had a sense of awareness about the world around them and their positioning in it. Their capacity for promoting change was evidenced in their discussions about social injustices and the negative impact their community. Through their participation in PEP, they were provided with capital that advantaged them. For example, to participate in the program, the students had to excel academically, and as a benefit for their participation, they received scholarships to fund their education in STEM education at the university and their progress to a degree was regularly monitored. As a part of a cohort, they received peer support, and the faculty of PEP provided mentoring and encouragement throughout their academic pursuit.

3.3 Aspiring to Make a Difference

As stated earlier, my scholarship is guided by a desire to use my knowledge and skills to shed light on social constraints imposed upon historically marginalized communities that obstruct their paths to achievement in STEM fields and careers. While I understood the vastness of this undertaking considering the overwhelming magnitude of the offenses, I at least want my voice to contribute to deep processing of issues for those who are negatively impacted by systemic inequities which is about more than reporting outcomes. It's about using my voice to explain the mission of social justice that is necessary for confronting and resolving inequities.

Navigating the expectations around what is required in academia and fulfilling a grounded purpose of advancing success in STEM education for my community of interest is my goal. While making a difference is not necessarily a requirement of scholarship, undoubtedly, a sole measure of success for scholarship is publishing outcomes. The expectation is not necessarily to solve problems, it is enough to simply report that the problems exists and the numbers of individuals who are impacted. Applying for funding counted; however, there is a risk if it is not funded, Then it could be considered a waste of time, particularly for junior faculty. I conclude that junior faculty from historically marginalized groups needs to be supported by someone who not only believes in their ability to succeed, but someone who is aware of the social implications of their identity and the potential impact of impostor syndrome. That would ease the pressure which—ironically—is applied to boost productivity. As a junior faculty, I felt the need to distance myself from the messaging in the atmosphere while I figured out who I was and what I was going to do. If I was going to be productive, I needed to manage the underlying pressure that would ultimately evaluate my work. This effort at transparency is in case there are junior faculty—like me—could find themselves in a similar situation, or to support those who are mentoring junior scholars from historically marginalized backgrounds or really from any background in academia.

Early in my academic career I realized that thinking about everything that needs to be done as a single undertaking could be overwhelming. De Clercq et al. (2019), refers to the overwhelmed state as time related work stress. De Clercq et al. (2019) defines time related work stress as "employees' sense that they have insufficient time to complete job-related tasks, due to excessive demands imposed on them by their work environment" (De Clercq et al., 2019, p. 1). Employees can feel overwhelmed by the magnitude of the work expectations and exhibit counterproductive work behaviors (CWB). Counterproductive work behaviors are negative work behaviors in response to stress, and while there are more intense negative work behaviors dependent upon personality types, some individuals may react by withdrawing and failing to complete tasks or engage in meaningful work activities, a form of paralysis. I was not aware of time related work stress and CWB as outcomes when I made a conscious decision to avoid situations that could feel over stressful. I felt that by shielding myself, I would be able to concentrate on my scholarship and even enjoy my work. My approach was to focus on one year at a time and one day at a time, to break tasks into manageable chunks to avoid becoming overwhelmed which could affect other aspects of my life. De Clercq et al. (2019) stated that employees who suffer from time related work stress are made to feel unhappy about their job situation and their futuristic opportunities. Rightfully so, academia has expectations with the promotion and tenure process, requiring certain milestones to be reached within a specific time frame. That is of premier importance. However, it would be helpful if information about promotion and tenure was shared by individuals who appreciate the challenges ahead and are confident in the junior faculty's potential to meet expectations.

In drowning out the background noise, I was able to focus on becoming a scholar. This did not mean that the messaging regarding the expectations went away, neither perspectives on the risks associated with the amount of time spent conducting qualitative research. While qualitative research was considered time consuming, I was able to focus on my scholarship. I found it meaningful to interview participants to gather their views on an intervention and to use that knowledge to make iterative advancements to the model. I also wanted to explain the theoretical underpinnings that supported deep understanding of challenges, which I consider pertinent to planning and implementing intervention. Examining a deficit without considering the historical, political, and sociocultural contexts contributing to the deficit is equivalent to putting a bandage on a wound without treating the underlying infection or causes. Treating the infection is a solution to the problem and can move the needle toward change, which is significant to having a positive impact on individuals from historically marginalized communities.

Morgan and Smircich (1980) in their discussion of the purposes of quantitative and qualitative research assert quantitative methodology as a way of examining the social world as a concrete structure. Strategies used to analyze data aim to position the world into what they refer to as "structured immobility" to limit human actions to a cause that follows a set of laws and principles that are not influenced by external factors. Quantitative research utilizes objective measurements for explaining the social world and assumes that laws govern the relationships between factors examined within a particular domain. According to Morgan and Smircich (1980), quantitative researchers are described as external observers, measuring what they see. Morgan and Smircich (1980), asserts that operating from a premise that reality is socially constructed requires an examination of the processes that created the reality. It is a subjective approach which examines the practices through which realities are communicated. Morgan and Smircich (1980) assert that qualitative research is the antithesis of examining the world as a concrete entity. The social world is more open-ended in that humans' realities are influenced by history, varying contexts, and varied forms of discourse. The qualitative researcher is not an external observer but examines occurrences within the phenomena to make sense of what is represented. Morgan and Smircich (1980) argue that the explanations of the choice between quantitative and qualitative research methodology are over simplified. They assert that the phenomena that is being observed, as well as the orientation of the researcher determines the forms of knowledge that is expected and produced.

Mir (2018) concurs with the notion that the quantitative and qualitative binary is oversimplified, referring to it as an "act of discursive violence." Mir (2018) reflects on a history when quantitative research emerged as the superior approach and was legitimized with high-level statistical analysis of large data samples. Qualitative methodology was seen as inferior to quantitative methodology and illegitimate. Qualitative research was heavily scrutinized, particularly with journals in terms of publications, and it was determined to be unhealthy for promotion and tenure decisions. Fast forwarding a century later, qualitative methodology experienced a surge in significance when it was noted that the most significant theories were processed and developed through qualitative analyses. Mir (2018) used the term "strategic essentialism," which was taken from postcolonial theory to point out the notion that the message is not to argue for the superiority of qualitative methodology, but to accept the binary and increase the value and respect for qualitative research. In the spirit of transcending the status of qualitative research, Mir (2018) makes three suggestions. The first is to resist analyzing data using quantitative thinking. The qualitative analysis process is not akin to large data samples, and researchers must stand vehemently behind decisions to explore the meanings and complexities of human experiences in social and cultural contexts. The second suggestion is to expand upon the explanation of ethics as more than "following the rules." In qualitative research, ethics is about more than protecting an individual. It is also about not sacrificing the stories of the oppressed for the powerful. The third and final suggestion is to be prepared for the opposition to qualitative research from institutions and resist it. Good research is authentic and follows principles that research evaluators value over career feasibility and idea conciliation.

I concur with Mir's (2018) beliefs that the research environment for qualitative studies is more accepting than it was years ago. I did not consciously adopt a mission to counter biases against qualitative research in my work environment, My approach was driven by the mission of my work and the process that was more viable for bringing it to fruition. As Morgan and Smircich (1980) asserted, it was my orientation as a researcher that dictated the methodology that I chose to fulfill work expectations. Over the years, I have employed mixed methods in project designs and found value in corroborating results through both lenses.

3.4 Conclusion

The underrepresentation of African American majors, in STEM, has been a major concern for decades. While throughout the years, there have been many initiatives devoted to increasing the numbers; however, meaningful change has been slow to materialize (National Science Board, 2018). African Americans are as likely to intend a STEM major upon entering college as Caucasian students; however, they persist at much lower percentages (Smyth & McArdle, 2004; Stipanovic & Woo, 2017). While increasing the numbers of individuals from underrepresented groups majoring in STEM fields is important, I am concerned about retention. I am concerned about the numbers of non-completers, who after meeting qualifications for acceptance face obstacles that impede their progress to achievement. According to Markle et al. (2022), there is a major disparity in the graduation rates between students from underrepresented groups and students from the majority group, particularly at PWIs. Markle et al. (2022) attribute this disparity to factors such as stereotype threat, impostor syndrome, and lack of social networks. These are all challenges that disproportionately affect underrepresented groups and are realities stemming from systemic disparities. Markle et al. (2022) assert that even though PWIs have placed a major emphasis on increasing the numbers of underrepresented individuals entering higher education, they have not provided the necessary support that will aid them in championing systemic obstacles that pose barriers to their achievement.

Nine years ago, I obtained funding to replicate a model for a university/public school collaborative high school robotics afterschool program that I co-developed and co-facilitated, the "STEM Through Robotics program" (STR) in a local school district near the university. The rationale for the STR model was to provide students with learning experiences that enhanced their self-efficacy, as well as their knowledge and skills, to prepare them for majoring in STEM fields. While it could be stated that a majority of the students enrolling in this original program came from resourced backgrounds, STR's success with a subset of students, who we later discovered had not planned to major in STEM fields, was significant. Students who expressed that they were not interested in STEM remained in this multiyear program throughout their time in high school. Findings revealed that some of these students enrolled in STR because it occurred after school, or included a competitive element, or was available as an evening elective course. Whatever their initial purpose for enrolling in the course, they persisted, with some of them indicating that they were considering changing their majors to a STEM field.

I believed that this program would have similar results if it was replicated in an under resourced public-school district in my home region and began setting the stage for its implementation. I discovered through initial talks with a district representative that approximately one percent of their graduates majored in STEM fields and a smaller percentage of them switched their majors while enrolled in college. It was decided that due to its close proximity to a Historically Black College and University (HBCU), this majority African American public-school district was an ideal setting considering the opportunity to collaborate with faculty in their computer science and engineering technology departments. Hence the development of a high school STEM outreach program, "Motivating Through Robotics" (MTR).. . Like the STR program, the program model was designed to influence the students' self-efficacy, but it also included a focus on their science identity in an effort to foster a sense of belonging in STEM. The MTR program was designed to provide physical and psychological access in a school community disparaged by systemic inequities.

> This program is more than a robotics program, it is a change in our mindset. Before being introduced to this program, we were not thinking like this. This program introduced us to new opportunities.

This statement was made by Marvin, an alumnus of MTR program which is now in its 10th year. In the robotics competition component of the program, there were very few African Americans scattered on teams throughout the regional event. At their first couple of events, they were the only all African American team. Since that time, there are at least two additional all African American teams out of about 60 teams participating in the event. This afterschool program exposed them to STEM activities that they would have not experienced. In their community, prior to this program, STEM activities were nonexistent. To influence a STEM culture, this program offered pipeline activities for students in grades 3–8 through summer enrichment activities. All of the program activities positioned them as problem solvers and contributors to thrive in a society that will become more technologically advanced and dependent in the future.

> We love you for giving us the opportunity to be a part of a world that we would have never known.

One of the public-school staff members made this statement at the end of the fifth robotics competition. She volunteered with the program throughout the five funded years and has remained committed to the program. Her son enrolled in the program as a ninth grader and participated in the program throughout his high school years. She indicated that she remained committed to the program because she recognized the program's benefit to the school district, and she wanted to sustain it. While the funding for the program ended after the fifth year of the grant, the school district has made earnest effort to sustain it beyond the funding period.

The capital that was afforded to the school community through this after-school program has resulted in several of them choosing STEM majors, whether at the community college or at the four-year institution level; however, there are challenges to persistence at the university level. Each year, alumni return to the program to volunteer and express the value of the program to their lives. On one occasion, a former student, without being prompted, shared her experiences with the students and encouraged them to become a leader. Quantitatively, statistical significance was found in students' attitude toward science which is corroborated in the qualitative data. While that is a sign of effectiveness, there is still much work to be done to equalize the playing field, especially in their academic preparation for majoring in STEM fields.

The relationships in the MTR program spring boarded my current collaborative project. Having met a very talented and knowledgeable instructor in the MTR program. I was inspired by his commitment and effectiveness in motivating the high school students in the MTR program to take responsibility their learning. In conversations with him I discovered that he paused his doctoral program before completing his dissertation. He shared life experiences that influenced his decision to put his doctoral program on hold. Through further discussion with colleagues at other HBCUs, I discovered that there were instructors with similar stories working at their institutions. It seemed logical that an initiative focused on increasing the numbers of doctoral degrees in STEM from historically underrepresented groups could be achieved through a focus on persistence and retention. In collaboration with colleagues at my university and HBCUs, we obtained funding for a project that focused on supporting historically underrepresented doctoral candidates in STEM in completing their doctoral degrees. The project was designed especially to support those whose work was in a state of paralysis, with priority given to individuals working as instructors at HBCUs. Markle et al. (2022), recommended structured mentoring networks for underrepresented students in higher education to support them with navigating challenges associated with their identity. This project focused on support systems integrating mentoring at varied levels, with an aim of supporting the candidates in resolving factors that disrupted their programs and prevented them from progressing. The focus on retention is out of respect and consideration of the immeasurable investment of time and resources that the underrepresented doctoral candidates had already committed to their doctoral programs. My goal, as is the objective of culturally relevant pedagogy, is to support them in refuting the narratives and understanding how these messages contribute to unequal playing fields for historically marginalized individuals in an effort to encourage agency (Mensah, 2021). I am inspired by the doctoral candidates who have used the project's resources to motivate them to push pass the hurdles that obstructed their achievement.

Five years ago, I was asked to present at a Black History luncheon for graduate students on my experiences as an African American female professor at a predominantly White institution. My presentation was entitled, "Drowning out the Background Noises: The Voices Only You Can Hear," the origin of the title of this book chapter. I not only presented on background noises related to my choosing a qualitative research approach for my work, but I also spoke about managing guilting messages around taking time for self-care, particularly carving out time for non-work-related activities. Considering they were graduate students, most of them doctoral students, I decided to share lessons learned, and my challenges with managing these struggles throughout the doctoral program and as a faculty member. There is always work that needs to be done and deadlines to meet. I believed that by hearing a faculty member admit to these challenges, it would expose this enigma and aid them in their decision-making processes. I encouraged them to think clearly about who they are and aspire to be.

In conclusion, I consider myself to be a sociocultural researcher, one who examines how social and cultural factors influence the beliefs and behaviors of individuals from historically marginalized groups' achievement in STEM. Examining these factors requires a thorough exploration on how biases evolved and how they operate as quiet often invisible forces beneath the surface. Unveiling these factors as a root cause for inequities aligns with the social justice mission that undergirds my work. As a sociocultural researcher, I examine how structural inequities within society marginalize certain ethnic groups, and how these individuals bear the stigma and the responsibility to refute them. I strategize models aimed at empowering those who are the most disadvantaged to realize the forces that contribute to their marginalization, and to support them in actualizing transformative agency (Giroux, 1981). In other words, I strive to support historically underrepresented groups in STEM in drowning out the background noise that muddles their achievement. My social justice missions is twofold, to expose the socially constructed barriers, and to translate this awareness into actionable strategies that confront the social inequities they produce.

Establishing a purpose in scholarship grounds the work. It goes beyond fulfilling a requirement; it is about being deeply engaged in work that is meaningful and purposeful. I discovered that affirming the value of your work is an important step toward establishing a scholarly identity. In my experiences at the university, while I was not immune to the background noise, I was able to create a space and embrace an identity that felt right for me. Despite the ongoing pressure to measure my progress against the achievements of others, I realized that the true value is found in the quality of the work I produce.

Ignoring the background noise enabled me to discover my path in academia, to define the meaningful impact that I wanted to achieve in my work. One word of caution from a committee member during my doctoral studies was the possibility of me becoming overwhelmed by the broad and complex work required in addressing and resolving inequities. He encouraged me to recognize my limitations and to appreciate the significance of the impact that I could make within my reach. He encouraged me to consider my contributions as significant and as a collective effort with others. I have reflected on that advice over the years, and with every achievement, I realize that there is even more work to be done. Thus, I resolved to work diligently in spaces available to me aiming for broader societal impact. The hope is that the lessons learned can benefit the efforts of others dedicated to addressing social injustices that impede the achievement of individuals from historically marginalized communities in STEM.

References

Anyon, J. (1980). Social class and the hidden curriculum of work. *The Journal of Education, 162*(1), 67–92.

Anyon, J. (1981). Social class and school knowledge. *Curriculum Inquiry, 11*(1), 3–42. https://doi.org/10.2307/1179509

Apple, M. W., & Weis, L. (1986). Seeing education relationally: The stratification of culture and people in the sociology of school knowledge. *The Journal of Education, 168*(1), 7–34.

Brand, B., Glasson, G., & Green, A. (2006). Sociocultural factors influencing students' learning in science and mathematics: An analysis of the perspectives of African American students. *School Science and Mathematics, 106*(5), 228–236.

Bourdieu, P. (1997). The forms of capital. In A. Halsey (Ed.), *Education: Culture, economy and society* (pp. 46–58). Oxford University Press.

Chubbuck, S. M. (2010). Individual and structural orientations in socially just teaching: Conceptualization, implementation, and collaborative effort. *Journal of Teacher Education, 61*(3), 197–210. https://doi.org/10.1177/0022487109359777

De Clercq, D., Haq, I. U., & Azeem, M. U. (2019). Time-related work stress and counterproductive work behavior. *Personnel Review, 48*(7), 1756–1781. https://doi.org/10.1108/PR-07-2018-0241

Foxx, K. (2021). Cultivating a sense of belonging: Black students at a predominantly white university. *Negro Educational Review, 72*(1–4), 107–129.

Freire, P., & Macedo, D. P. (1987). *Literacy: Reading the word & the world* (Ser. Critical studies in education series). Bergin & Garvey.

Giroux, H. A. (1981). Hegemony, resistance, and the paradox of educational reform. *Interchange: A Quarterly Review of Education, 12*(2–3), 3–26. https://doi.org/10.1007/BF01192105

Giroux, H. A. (1992). *Border crossings: Cultural workers and the politics of education.* Routledge.

Hussain, M., & Jones, J. M. (2021). Discrimination, diversity, and sense of belonging: Experiences of students of color. *Journal of Diversity in Higher Education, 14*(1), 63–71.

Kohl, H. (1994). I won't learn from you: The role of assent in learning. *Harvard Educational Review, 64*(2), 233–233.

Kohl, H. (1995). I won't learn from you and other thoughts on creative maladjustment, by H. Kohl. *Communication Education, 45*(3), 263–264.

Markle, R. S., Williams, T. M., Williams, K. S., deGravelles, K. H., Bagayoko, D., & Warner, I. M. (2022). Supporting historically underrepresented groups in STEM higher education: The promise of structured mentoring networks. *Frontiers in Education, 7.* https://doi.org/10.3389/feduc.2022.674669

Mensah, F. M. (2021). Guest editorial. *Science and Children, 58*(4), 10–13. https://doi.org/10.1080/00368148.2021.12291647

Mir, R. (2018). Embracing qualitative research: An act of strategic essentialism. *Qualitative Research in Organizations and Management: An International Journal, 13*(4), 306–314. https://doi.org/10.1108/QROM-09-2018-1680

Morgan, G., & Smircich, L. (1980). The case for qualitative research. *The Academy of Management Review, 5*(4), 491–500.

National Science Board. (2018). *Science and Engineering Indicators 2018.* National Science Foundation (NSB-2018-1). https://www.nsf.gov/statistics/2018/nsb20181/report/sections/higher-education-in-science-and-engineering/highlights

Smyth, F. L., & McArdle, J. J. (2004). Ethnic and gender differences in science graduation at selective colleges with implications for admission policy and college choice. *Research in Higher Education, 45*(4), 353–381.

Stipanovic, N., & Woo, H. (2017). Understanding African American Students' experiences in STEM education: An ecological systems approach. *The Career Development Quarterly, 65*(3), 192–206. https://doi.org/10.1002/cdq.12092

Chapter 4
Right Where I Belong: Fighting for the Space in Science That Everyone Else Is Given Without Question

Terrance Burgess

4.1 Introduction

"For all those who have to fight for the respect that everyone else is given without question" (Jemisin, 2015). In N. K. Jemisin's *The Fifth Season* (2015), she begins her novel with a dedication to "those who have to fight." This notion of "fighting" will be explored throughout the course of this chapter. On the pages succeeding the dedication, Jemisin orients the reader to the context of her story with a map of the land she refers to as "The Stillness." This land resembles what we may geologically refer to as a supercontinent and is comprised of two tectonic plates, the Minimal plate and the Maximal plate. The map includes some of the topographical features one might expect to find on a map of an actual place (e.g., mountains, bodies of water, etc.), complete with a scale demarcated in 1000-mile increments. Jemisin further describes The Stillness as a tectonically active place, where its citizens' existence is predicated on a perpetual fear of the next geologic event. To describe a typical day, Jemisin states,

> Here is the Stillness, which is not even still on a good day…The earth is good at healing itself. This wound will scab over quickly in geologic terms, and then the cleansing ocean will follow its line to bisect the Stillness into two lands. Until this happens, however, the wound will fester with not only heat but gas and gritty, dark ash—enough to choke off the sky across most of the Stillness's face within a few weeks. Plants everywhere will die, and the animals that depend on them will starve, and the animals that eat those will starve. Winter will come early, and hard, and it will last a long, long time. It *will* end, of course, like every winter does, and then the world will return to its old self. (pp. 7–8, emphasis in original)

This vivid articulation of the Stillness specifically resonates with me, as I, a Black man and geologist, chose to study a discipline that was not intended for my

---

T. Burgess (✉)
Department of Teacher Education, Michigan State University, East Lansing, MI, USA
e-mail: tburgess@msu.edu

M. M. Atwater and B. A. Garvin (eds.), *Creating and Maintaining a Home in Science Education Research*, Contemporary Trends and Issues in Science Education 65,
https://doi.org/10.1007/978-3-031-99382-4_4

pursuit. Much like the Stillness, my development in STEM was constantly thwarted by cycles of destruction and regrowth, leading to my urge to fight. This fight, while not physical, encompasses many of the challenges highlighted throughout critical science education literature. I detail them below and offer a reimagination of what a fruitful science experience that centers one's belonging and promotes their mattering might look like.

4.2 Pre-college Science (In)Experiences

4.2.1 Elementary Years

Aligning with national data regarding the current state of the minimal science instruction (Banilower et al., 2018), my elementary school years were practically devoid of science. Although occurring three decades ago, I distinctly recall science experiences being comprised of an alignment with the seasons of the year rather than pursuing answers to my most burning questions. For example, as school started in August and fall was on the horizon, our first 'science' unit was comprised of a month-long study of apples. We did activities such as making apple sauce, bisecting an apple to reveal the star-shaped seed pattern within the core, and completing coloring pages with a cartoon-illustrated apple with a worm sticking out of it. This was my science. However, I was interested in more than just apples; I had much bigger questions. As a child who grew up in rural North Carolina, I was responsible for helping my father to load firewood onto the back porch to keep the house warm during the winter. He would split logs with an axe while we, the children, stood back and watched. This was fascinating to me because the wood chips reminded me of a snack I had come to eat regularly—saltine crackers. Childhood me was convinced that saltine crackers must be composed of wood pulp because how else were they so crunchy? At the time, I could not envision a dough being formed, pressed, cut into squares, and baked; saltine crackers resembled wood chips and were hard like wood. That made sense to me and was how I had come to develop an initial interest in science.

Given this intriguing wondering, I would later plan excursions into the woods behind my house, where I wanted to simply know more about my surroundings and how things worked. I would spend hours exploring, never noticing how long I had been gone from home. These excursions involved developing a sense of direction to prevent getting lost along with how to identify poisonous plants to maintain safety. I also learned about various rocks and spent hours searching for arrowheads. This was critical science knowledge that I had developed well before beginning school.

Another distinct memory of my childhood science was the question of how to model tornadoes at home. I vividly remember my kindergarten (and most influential) teacher, Ms. Stewart, sitting with the class around the carpet and asking us if we wanted to see a tornado. "We can't see a tornado because we are not outside," I replied. Then, suddenly, she removed these two connected soda bottles from beneath her desk. One of the bottles was partially filled with soap and water, while the other bottle was inverted and connected via a plastic collar. She then inverted the bottom bottle as though initiating the countdown of sand in an hourglass. As she flipped the bottle, she began to rotate the connecting collar, creating a waterspout within the bottle. My mind was immediately blown. How could she create a tornado in a bottle?! Why did it seem so simple? I had to know the answer to these questions for myself.

To explore my own answers to these questions, I begged my parents to buy two 2-L sodas from the nearby gas station, not telling them that I was going to recreate the model my teacher showed us during class. After my siblings and I drank the bottles' contents, I began my investigation. As I witnessed in class, I filled one of the bottles halfway with water and soap and, in lieu of a plastic collar, used duct tape to affix the top inverted bottle to the bottom bottle. I turned the bottle system just as I had watched Ms. Stewart do earlier, and then, to my disappointment, the water spilled from between the loosely connected strands of tape. I had come to reason that the plastic collar in Ms. Stewart's tornado demonstration must have played a crucial role in the creation of her model. However, not discouraged, I decided to see what else I could make with the plastic bottles, sparking my interest in rockets.

Moving my "lab" from the kitchen to the backyard, I decided I would take the bottles and create a rocket system. From television, I knew rockets were launched from a launch pad, yet I had no idea how I would get my hands on such a device. Then, it occurred to me that I could create my own. Taking one of our sharpest kitchen knives, I began to saw the top quarter of the bottle so that I could use it as a funnel to rest the other bottle on as it was launched into the yard. In typical child fashion, the knife was unsteady and slipped from the bottle and into my finger as it moved back and forth. As I looked down at my hand, I noticed a large gash on the joint of my middle finger. Rather than crying or telling my parents (I knew I would be in a lot of trouble), I decided to dress the wound myself and abandoned my interest in rockets for an initial interest in medicine. Even today, decades later, I carry the scar from this adventure. Pursuing a career in medicine meant I would get to find answers to my most burning questions. Given my wounded hand, I wanted to know how it was that the body seemed to automatically know how to heal itself and how I could use this knowledge to help others.

While neither explicit nor literal fights, my elementary years were filled with times in which I had to fight against my science (in)experiences to create opportunities for myself to better understand the world around me. I had questions that needed answers—deep wonderings that I would stop at nothing to explore further. I engaged in these explorations despite what was not happening at school, and unbeknownst to my parents, I felt like a scientist on the brink of a breakthrough. The absence of science was understood rather than challenged, leading me to believe it to be normal. However, it would be years later before I realized the importance of cultivating a multi-dimensional science identity as a young child (Burgess, 2020). In fact, during my high school years, I further realized the significance of this nurturing.

4.2.2 Middle School Years

Throughout my middle years, science was sporadic and primarily taught by long-term substitute teachers. However, there is one science memory that stands out to me. In eighth grade, we were assigned a project in which we had to create a three-dimensional model of a selected biome. I elected to create a model of a tundra, as I had never heard of this biome before. I was fascinated by the permafrost and the freezing temperatures, wondering how it was possible for life to exist in such a place. From my studies, I knew that the permafrost trapped methane and that warming temperatures could wreak havoc on our climate. Armed with this information, I persuaded my parents to purchase glue sticks and miniature animals and I began constructing my model tundra. I distinctly remember gathering dead grasses, twigs, moss, and pebbles to create my biome. We were given about three weeks to complete this project, and at the end of our ecology unit, we were expected to bring our biome models to class and present our findings.

Now, a few weeks later, I excitedly dragged my oversize posterboard and other materials to school to finish assembling my masterpiece in class. After getting to class, I excitedly began putting the finishing touch on my poster model. With its beautiful backdrop and 3-dimensional elements, I thought it was perfect. Our teacher told us that we would have the chance to present after lunch and I could not wait. However, during lunch, several students in class were involved in a fight and because of this, our teacher decided to postpone our presentations. After returning from lunch, she yelled "Since you all don't know how to behave, we won't do our presentations today!" to which the class responded in groans. Instead, we were forced to sit silently and complete worksheet pages from the textbook. This was devastating to me as I had spent so much time on something that I was beyond proud of. Unfortunately, this was an all-too-frequent occurrence. Science time was always the first to be sacrificed in favor of more "important" subjects, and textbooks were used to manage bodies and behaviors that were illustrative of children simply being children. While the joy of science was actively being stripped from me, I was more determined than ever to fight for more science.

4.2.3 High School Years

I was very excited to enter high school and get to formally—and consistently—study science. However, this excitement was fleeting. As the standard course of study, we were required to take a life science, an earth science, and a physical science course for graduation. Typically, those interested in pursuing college also took an additional physical science course (i.e., physics), and although the course was offered, having a teacher trained in the subject was as unpredictable as an earthquake in The Stillness. In many ways, science felt like a privilege rather than a right, leading to yet another instance of having to fight. To paint a more comprehensive picture of my science journey, I detail a few experiences from these formative years.

Earth Science. Ninth-grade earth science was my first foray into what I would eventually study in college. However, this was nowhere near a foundational experience. My earth science course was chaotic; we were a group of 30 teenagers piled into an old lab classroom surrounded by textbooks and stacks of copier paper. I remember the blue hardback textbook very, very well. For all intents and purposes, this textbook was my teacher. In other words, my actual teacher sat at her desk and directed us to complete packets of worksheets where we would have to search the text for the answer to the blanks on the page. While I typically found word search puzzles entertaining, this was grueling. It was nothing like I thought high school science would be, and it would be years before I realized this was unacceptable. I envisioned a classroom with Bunsen burners, lab coats, and mysterious liquids waiting to be combined with one another. Instead, we received a very traditional experience. For example, during the academic year, our teacher only gave one lecture, which happened to be on plate tectonics. To model this phenomenon, she would use her hands to represent continental plates and would move her hands in various directions to illustrate plate motion. From this lecture, I left believing that the earth's plates rested on magma, much like a boat on water. Like my understanding of the relationship between saltine crackers and wood, this made sense to me. There was no need to question this knowledge—nor my teacher's methods. Sure, it could be more engaging, but at least we were concentrating on science, right? This is not to besmirch the character of my teacher. I am sure she meant well; however, there was a stark difference between meaning well and teaching well, and unfortunately, the latter suffered tremendously, shaping my early conceptualizations of what science was and could be.

Life Science. During tenth grade, I took a biology class, the only state-tested science course at the time. Because of this test, the stakes for success were high. The test was partially responsible for the assessment of the school as a whole, and if you failed to pass the test, you automatically failed the course and effectively failed the grade. In my case, the class was taught by a biologist, so I was beyond thrilled. We were in another lab-style classroom, and on the first day of school, our teacher broke the news to the class: "*This year, there's no state test. The state is administering a field test, so none of the items will count for you. So, all I ask is that you behave, and you'll pass the class.*" As a result, I learned to play several card games in Biology, yet knew nothing about the discipline. Our days were filled with wondering about the possibilities of a real biology course. Here, yet again, was another missed opportunity. At one point in the semester, we finally convinced our teacher to allow us a chance to do some hands-on and authentic biology. This resulted in us dissecting a fetal pig, which, for me, was exciting as I had not yet been exposed to any "true science." I fondly remember this assignment being loosely structured, resulting in a free-for-all with the dissection materials. We were expected to complete a worksheet but had no discussion of what we learned. Sadly, I cannot recall much else from this time, as there were no memorable moments that spawned any kind of positive feelings toward science. In fact, before becoming a teacher, I had grown to fear biology.

Senior Year Wonderings. Although my high school science experiences left much to be desired, I continued my wonderings about the world, thus maintaining my curiosity in science. By this time, I was driving to school and remember having this distinct question: *Why is there frost on the windshield of my car during the fall and winter if the surrounding air temperature is above freezing?* I had come to reason with myself that, somehow, the surface affected by the freezing water had become colder than the surrounding air, allowing the water to freeze upon its surface, and while this made total sense to me, my English teacher was less than impressed with my thinking. Rather than applaud and encourage my thinking, she decided to make this a "teachable moment" for the rest of the class at my expense. She berated my ideas and threatened to call home to my parents if I kept talking, yet another instance of my fighting. In fact, while this interest (like so many others) was self-motivated, it also serves as a case in which I made several bids for my science identity to be recognized which could have greatly reshaped the course of my science trajectory.

4.2.4 College Science (In)Experiences

My college science experiences were a bit different, yet the results were very similar. For example, I became a geologist purely by happenstance. At the time, I needed one additional course to be considered a full-time student, and a first-year seminar on the geologic history of North Carolina was the first available course, so I enrolled. This course became one of my favorites, leading me to continue taking geology courses until I had accumulated enough credit hours where it simply made sense to pursue the major. Through this geology seminar, I was encouraged to continue asking questions and told by my professor that I had a "natural knack" for geology, illustrating the power of teacher affirmations. I would always grapple with questions that were beyond the scope of the course yet were very interesting to me as a student. This was a course that I looked forward to, and I somehow thought the rest of the major courses would be the same experience. Sadly, I was mistaken.

The subsequent 3 years within the major were tough. I remember spending hours on homework and lab assignments in classes, wondering how it was that everyone else seemed to have this insider knowledge about geology while I was struggling to keep up! I vividly remember sitting in classes hoping that I would somehow absorb the information being spewed in my direction much like a dry sponge in water. I—as the only Black major—would often attend office hours where professors would not know my name despite our small department of 30 undergraduate majors. Whenever I decided to take courses outside of the department, I was immediately contacted by advising for fear that I had left the major, causing the department to lose its' claim of diversity. So, clearly there was some need for faculty to care. However, this care was only extended when it served their (i.e., the department's) purposes.

Additionally, I studied geology in a building that was named after a slaveholder who believed that Blacks were 'morally and mentally inferior.' During this time, I blamed myself for what I thought were my own 'academic shortcomings' and had not yet realized that my experiences were emblematic of a greater system in which young men like me become perpetuators of said cycle. This system was one in which fighting was absolutely necessary, and unlike my high school experiences, I was unaware of the systemic implications. However, after studying geology, these (in)experiences became ones which would eventually fuel my personal interests in education.

4.2.5 Becoming a Science Teacher

Upon graduating college, I decided I was no longer interested in pursuing a career in medicine. Instead, given my struggles with education, I decided to become what I so desperately wanted to experience growing up: a 'good' science teacher. I had recently completed a program in which my department partnered with the school of education allowing me to take the state licensing exam and complete student teaching, and I would graduate with a comprehensive science (grades 9–12) teaching license. To me, a good science teacher was one who allowed students the space to experience true science learning for themselves. Science was an *action* rather a *thing* and was filled with wonder and excitement. Despite this action, scientific knowledge still resided squarely within the canon. Because of this, I emulated many of the pedagogical decisions that made science an unenjoyable experience for me. For example, in Advanced Placement (AP) Environmental Science, a course I was quite fond of, I firmly believed that college-level science began and ended with an understanding of the discourse used to communicate ideas. Thus, I would have students spend time at the beginning of each unit "learning" the unit vocabulary, as this was what I thought offered them access into the discipline.

In physical science, a hybrid course between general physics and chemistry, I taught students who were practically the same age as I was at the time. While I had figured out what I thought 'good' science teaching was, I quickly added another variable to that equation—*classroom management*. This course was a graduation requirement for all, but lower-tracked students were the only ones who occupied this course. So, I quickly thought that younger me had to have physical control over my classroom. Because of my age, I had to let the students in my class know that I was to be taken seriously, completely missing the opportunity to truly connect with them. This was the only science class in which all the students looked like me, yet I relied upon the same carceral tactics I grew up in to "maintain control" of the classroom. We seldom did labs, as I was quickly warned by other teachers that "these students could not handle" lab activities due to their behavior. Thus, my science class closely resembled the same science that I experienced growing up—relying on the completion of worksheets.

In my honors—and all white—Biology class, I relied heavily upon the textbook and previously administered state tests to guide my instruction. Contrary to my own experiences in biology, my students would at least experience regular opportunities to engage in hands-on learning. However, they would also spend countless hours preparing for the state test, which lessened the authenticity of the course. I also had to constantly prove my knowledge to the affluent White students taking my advanced courses. If they received a grade that they felt was less than desirable, I would get questions such as "Mr. Burgess, what page is this answer on in the textbook?" Or I would receive emails from students and their parents pleading for extra credit to receive a higher grade in the course. In this case, I was quickly reminded of why I spent so much time fighting only to have to engage in a different kind of fight.

Near the end of each academic year, students would come around to each of their teachers asking us to initial their proposed schedules for the next academic year. This felt very similar to a "get to know you" bingo where you walk around and learn as much about your peers as possible. However, in this case, students were not learning about one another. Rather, they were learning about how their teachers perceived their abilities. This was the most blatant form of tracking, and I could not believe that so much power resided with one teacher who could change a child's entire academic trajectory. To this end, I want to tell a brief story about a brilliant young person we shall call Simone.

Simone, a Black girl, was tracked into my earth science class at the beginning of her ninth-grade year. However, Simone was one of my most vocal students and was clearly not finding the course material to be challenging enough. One day during the second week of school, I decided to ask Simone why she was not in my honors Biology course. "Because I'm not *that* smart, Mr. Burgess." This elicited a feeling that was all too familiar for me. "Simone, why do you think that?" I asked. "Because I'm not in any other honors classes, plus I struggle in my English class" she replied. It so happens that students are expected to have a certain eighth grade reading score to take biology as a ninth grader. Or, at least, this is the message I received. However, not satisfied, I decided to talk with Simone's parents. After explaining the situation and letting them know that I thought Simone was more than ready for Biology, they agreed and contacted the school counselor to move her into my biology class. Simone was brilliant and by encouraging her to take Biology early on she could take Advanced Placement (AP) courses in the future. This was a system in which students could work around tracking. It is also demonstrative of my having to fight for others to have access to what they were entitled to.

4.2.6 On Radical Hope & Belonging

Carl Grant (2020) defines *radical hope* as "hope that develops out of suffering and mourning, the destruction of a way of life in the face of anxiety and uncertainty" (p. 67). This definition could not be a more fitting description of my experiences in STEM. These times were trying and sometimes harmful, yet I persisted. Why? This is a question that I do not quite know how to answer, as I constantly try to recall what was so special about *me* that allowed me to "make it through." The truth is, I was somehow able to navigate the system that is inaccessible to so many of us, largely through acts of fighting for what others were also entitled to. I argue that while Grant defines radical hope through the lens of suffering, I want to encourage others to think about this suffering as a means to an end. We should not have to endure suffering for mere access to one's civil rights (Tate, 2001). Thus, we continue forward to claim our legacy.

Rather than situating this claim squarely in the space of radical hope, I urge those who share my STEM experiences to think of their presence as emblematic of their *radical belonging* (Louie et al., 2022) in STEM. In this case, radical belonging exists within the intersections of social belonging (the acceptance of people as fully human), academic belonging (school's acceptance of learning that is agentic and anti-hierarchical), and democratic belonging ("[one's] knowledge, skills, dispositions, and structures that it takes for a learning community to work for justice at a systemic level" [p. 12]). Additionally, radical belonging is existing within those spaces not meant for you and subsequently thriving. It is more than persisting, as this implies suffering. In reality, thriving encompasses making space for yourself to unapologetically exist and reap the benefits of joy, peace, and purpose. So, I ask: what would radical belonging have looked like throughout my STEM trajectory? Next, I reimagine my experiences within these spaces of belonging.

Social Belonging throughout College. Louie et al. (2022) view social belonging much like Carey (2019) views *comprehensive mattering*, which argues that young Black boys and young men "matter robustly" (p. 383). During my college years, belonging socially in geology would have looked like an acceptance of the funds of knowledge (Moll et al., 2006) curated throughout my formative years. Social belonging would have also accounted for the ways in which my limited formal science experiences shaped these funds, and that, systemically speaking, my opportunities to engage with high-quality science learning looked vastly different from my White counterparts. This is not intended to evoke sympathy as much as it is intended to evoke understanding. It would have looked like professors knowing my name and sharing a genuine interest in who I was as a human being while also ensuring that I felt welcomed within that space and that my ideas and wonderings mattered comprehensively (Carey, 2019).

Social belonging holds significant implications for learning, in which Louie et al. (2022) argued that "it shapes people's willingness to ask questions, share inchoate ideas, receive feedback, and be vulnerable with one another" (p. 9). While one professor appreciated and encouraged my question-asking, therefore promoting my social belonging, others viewed this act as a limitation of my knowledge, thus preventing me from being vulnerable altogether. In other words, my professor implicitly communicated through his (in)actions that I did not belong socially in his course. Had I belonged fully and socially within this space, no single question would have been too simple. Additionally, all questions would have been welcomed, with instructors anticipating which ideas students would be most likely to struggle with, developing an instructional course of action to meet these students' needs. Instead, my questions led to increased frustration among some of my instructors and counterparts leading to feelings of being intellectually inferior. Vulnerability was masked under the veil of a deep conceptual understanding (i.e., nodding in agreement with my counterparts despite not knowing what I was nodding for). Social belonging would have meant that throughout my entire STEM learning trajectory, I belonged academically.

Academic Belonging throughout Schooling. Academic belonging argues that schools and other educational institutions cannot truly be antiracist if they do not center academic belonging for Students of Color (Louie et al., 2022). Further, academic belonging "envision[s] learning that is agentic, responsive, collaborative, and anti-hierarchical" (p. 9).

Agency. To truly belong academically throughout my STEM schooling experiences, I should have been viewed as a being who was able to create knowledge. Rather than shutting my ideas down, my high school English teacher could have promoted my academic belonging by simply asking me to "say more." *How did you reason that the surface of your car had reached sub-freezing temperatures? How might you investigate this claim?* Arguably, these are questions (or versions of them) are also applicable to elementary-aged children, meaning that my curiosities about tornadoes and rockets would have replaced the seasonal—and perhaps empty—learning about apples had they been properly tapped into. These conceptualizations of science would have greatly impacted the quality of my future science instruction, as it would have given my parents a point to leverage in their advocacy for my science learning. There is current scholarship that demands that other ways of scientific knowledge be solicited and valued (e.g., Gunckel et al., 2024; National Academies of Sciences Engineering, and Medicine [NASEM], 2021), yet my *personal* experiences nearly three decades ago recommended the same, but these calls went unheard. In so many ways, my experiences were informing research that had not yet existed, reiterating the power of listening to children. Academic belonging would have ensured that my ways of being were acknowledged and that my teachers were responsive to them.

Responsiveness. Academic belonging attends to the myriad identities that encompass a person. More specifically, it "involves teaching in ways that connect to, value, and build on the backgrounds and goals of [S]tudents of [C]olor, their families, and their communities, decentering whiteness as the axis around which all learning must revolve" (Louie et al., 2022, p. 10). Coupled with centering my ways of knowing and doing science, academic belonging means featuring scientists throughout curriculum materials that look like me. It means that my ways of speaking and doing science are just as valued in earth science as they are at home. It means being able to teach in ways that connect with me beyond simply centering your own professional and personal research interests in undergraduate geology courses. In my teaching experiences, responsiveness would have meant ensuring that my students felt connected to what we were learning beyond merely meeting the needs of an assessment. Their ideas—rather than their complicity—would have been both centered and valued within the space such that they saw themselves as members of the science community.

Anti-Hierarchical and Collaborative. Academic belonging effectively "levels the playing field" between the instructor and the student. In other words, it reduces (if not eliminates) the power structures that exist within the classroom such that both students and teachers can collaborate in ways that co-construct knowledge. For me, academically belonging would have meant that my English teacher would work with me to develop an understanding of the frost on the windshield. It would mean that my professors knew my name and that I, as a classroom teacher, worked with students to develop shared knowledge about environmental science. Instead, I prioritized academic discourse that alienated some while others who were familiar with the game of science terminology recall thrived, resulting in the continuation of harm in the name of 'true' science instruction. I am clear to implicate my own teaching, as it is reflective of what Paris and Alim (2014) would call a "loving critique." Having been indoctrinated into the very systems that I thought I was disrupting made me a willing participant in perpetuating the status quo. It was not until I was forced to reconcile my science learning with my science teaching that this became apparent. Such a reconciliation is ever important for science instructors, as it allows us to do the work of equity (Burgess & Patterson Williams, 2022). In doing so, we commit to acknowledging the racialized nature of schooling (McKinney de Royston & Nasir, 2017), which allows us to understand and name that science learning itself is a racialized endeavor. It also requires us to acknowledge the harm we have done in the name of science while committing to doing no new harm (Morton et al., 2022). Through these acknowledgements we can actively work to disrupt this system for the equitable distribution of power. This disruption only occurs through the promotion of academic belonging.

Democratic Belonging in the Classroom. Democratic belonging is defined as "encompassing the knowledge, skills, dispositions, and structures that it takes for a learning community to work for justice at a systemic level" (Louie et al., 2022, p. 12). Returning to my time as a high school teacher, for me to truly promote democratic belonging, it required that I acknowledge the racialized nature of science teaching and learning while thinking about science teaching and learning from a systems perspective. Stated differently, I should have viewed the science that I was teaching from the Western/White perspective in which it was situated (Mutegi, 2013). Further, democratic belonging involves the raising of a critical consciousness while promoting participatory action (Louie et al., 2022).

In practice, a clear example is akin to what Daniel Morales-Doyle and Alejandra Frausto (2021) refer to as *youth participatory science*. The YPS framework combines Youth Participatory Action Research (YPAR) with citizen, or *street* science (Corburn, 2005) to select and deeply engage in science learning within a community-based context. A YPS learning cycle consists of several components: *define the social justice science issue (SJSI), apply a scientific lens, plan & conduct an investigation, analyze data & assess learning,* and *reflect, disseminate, act*. The framework actively centers communities and students who have been historically marginalized through issues rooted in systemic racism (e.g., redlining, environmental racism, etc.) and leverages scientific knowledge as the catalyst for change. Described as "a specific form of street science, which democratizes the tools of science in a way that values the wisdom and understandings that exist within communities marginalized by racism and economic dispossession" (p. 2), the framework is unique as it acknowledges the cultural capital and knowledge inherent within the community rather than treating it as separate from the learning that is expected to occur within classrooms. Throughout this instructional cycle, it is important to note that although these steps are cyclical, they can occur in various arrangements. For example, after applying a scientific lens, one may be able to analyze data and assess learning. Contrarily, after applying a scientific lens, one may decide to revise and conduct the investigation again to generate new data to analyze in pursuit of a solution. Together, democratic belonging complements social and academic belonging such that they are inextricably linked. Promoting these efforts as a classroom teacher would have greatly improved my praxis.

4.3 Discussion

What is the significance of recounting this STEM journey? Perhaps it is to inspire others who share a similar story. Perhaps it is to encourage those who look like me to keep going. Perhaps it is all these things. Radical belonging requires that we face these questions, as this is the only way in which we can truly thrive. Jemisin's (2015) text prepares us for an end. As with any end comes new beginnings that allow us to chart a new path forward. Much like the ecological succession that occurs after a volcanic eruption, there is new life that arises from the ashes. In my experiences, this life is best embodied in my current role as a science educator. While I do not currently practice geology, I *am* a scientist. The students I teach *are* scientists and through my pedagogical decisions, I strive to enact radical belonging. As a science teacher educator, I ensure that the stories of those who are typically ignored or overshadowed in science make their way into my classroom and that future teachers understand that part of the educational debt (Ladson-Billings, 2006) we owe children is ensuring they have rich and humanizing science experiences. My childhood experiences are what have shaped me, allowing me to narrate my identity (Kane, 2016). Further, they continue to shape my scholarship. It is through belonging that I can make these connections and advocate for a more humanizing approach to science teaching and learning.

Throughout the k-16 classroom, it is imperative that we nurture the budding science identities of learners, which, indeed, is an emotional endeavor (Avraamidou, 2020). They have vastly different ways in which they make sense of the world, and this is worth honoring. This is particularly important during the elementary grades, where children receive messages (whether implicit or explicit) that science is reserved for a particular type of person (i.e., white males). This narrative must be disrupted so that children of Color (in particular) can see themselves as belonging and valid contributors to the canon. Much like Kimmerer's (2013) depiction of asters and goldenrods, we have cultural traditions that allow us to name it as science. This science is survival. It allows one to know what and how much to take from the land, along with its implications. Through Western science, our surroundings are promoted as a beast to be tamed rather than an entity to be understood. Humans are regarded as superior beings when, in fact, we have been preceded by so much life, which then raises the question—*Where do humans truly belong in science*?

4.4 Concluding Thoughts

Throughout my STEM experiences, I have grown to have a deeper love for science and *myself* as a *scientist*. The institutional norms of teaching science the way we learned it are antiquated and dehumanizing, while contemporary notions center children's funds of knowledge, their language(s), and their very being. In other words, contemporary science supports radical belonging. Through this belonging, science takes on a different meaning, as the epistemological foci of science itself is broadened to include myriad cultures and perspectives. It is only through this belonging that we begin to shift the narrative about what counts as science, who is entitled to science, and whose ideas get valued.

In the science classroom, this presents as acknowledging and honoring one's social, academic, and democratic belonging. Within the elementary classroom, this means viewing science as a discipline in which children and their ideas matter greatly and are reflective of the scientific enterprise. In high school, it means taking students' wonderings seriously and supporting them in developing deep conceptual understanding. In college, this means encouraging one's curiosity while also understanding that while no two student's journeys through STEM are the same, their knowledge is equally as valid. As a science teacher, it means challenging the antiquated ways of learning in favor of a more humanizing and student-centered approach that acknowledges the myriad science epistemologies of its learners. Given the Stillness' unpredictability, it seems that one's science experiences must mirror this, which is not the case. As a science teacher educator, I ensure that my students view the future students in their classrooms as belonging radically in science. Understanding that this may be the only time in which I can fully advocate for science instruction, I not only hold on to Grant's (2020) radical hope, but I model what rich and just science experiences are possible for children. We must fight for what others are entitled to and do so unapologetically. Only then is the fight worth it.

References

Avraamidou, L. (2020). Science identity as a landscape of becoming: Rethinking recognition and emotions through an intersectionality lens. *Cultural Studies of Science Education, 15*(2), 323–345.

Banilower, E. R., Smith, P. S., Malzahn, K. A., Plumley, C. L., Gordon, E. M., & Hayes, M. L. (2018). *Report of the 2018 NSSME+*. Horizon Research Inc.

Burgess, T. (2020). *"Because I did what she told me to do." Understanding the identity work of elementary students of color in the science classroom and how a teacher's positionality impacts student learning* [Doctoral dissertation]. Syracuse University. ProQuest.

Burgess, T., & Patterson Williams, A. (2022). Utilizing theory to elucidate the work of creating equity for transformation within the science classroom. *Science Education, 106*(5), 1071–1083.

Carey, R. L. (2019). Imagining the comprehensive mattering of Black boys and young men in society and schools: Toward a new approach. *Harvard Educational Review, 89*(3), 370–396.

Corburn, J. (2005). *Street science: Community knowledge and environmental health justice.* MIT Press.

Grant, C. A. (2020). Radical hope, education and humanity. In *The future is black* (pp. 65–71). Routledge.
Gunckel, K. L., Davis, E. A., & Bautista, J. (2024). Moving beyond equity-as-access: Expanding what counts as science in the elementary classroom. *Science and Children, 61*(1), 32–37. https://doi.org/10.1080/00368148.2023.2292392
Jemisin, N. K. (2015). *The fifth season.* Orbit.
Kane, J. M. (2016). Young African American boys narrating identities in science. *Journal of Research in Science Teaching, 53*(1), 95–118. https://doi.org/10.1002/tea.21247
Kimmerer, R. W. (2013). *Braiding sweetgrass: Indigenous wisdom, scientific knowledge and the teachings of plants.* Milkweed Editions.
Ladson-Billings, G. (2006). From the achievement gap to the education debt: Understanding achievement in US schools. *Educational Researcher, 35*(7), 3–12. https://doi.org/10.3102/0013189X035007003
Louie, N., Berland, L., Roeker, L., Nichols, K., Pacheco, M., & Grant, C. (2022). Toward radical belonging: Envisioning antiracist learning communities. *Race Ethnicity and Education*, 1–21. https://doi.org/10.1080/13613324.2022.2106879
McKinney de Royston, M., & Nasir, N. S. (2017). Racialized learning ecologies: Understanding race as a key feature of learning and developmental processes in schools. In N. Budwig, E. Turiel, & P. D. Zelazo (Eds.), *New perspectives on human development* (pp. 530–573). Cambridge University Press.
Moll, L., Amanti, C., Neff, D., & Gonzalez, N. (2006). Funds of knowledge for teaching: Using a qualitative approach to connect homes and classrooms. In *Funds of knowledge* (pp. 71–87). Routledge.
Morales-Doyle, D., & Frausto, A. (2021). Youth participatory science: A grassroots science curriculum framework. *Educational Action Research, 29*(1), 60–78.
Morton, T. R., Miles, M. L., Roby, R. S., & Ortiz, N. A. (2022). "All we Wanna do is be free": Advocating for Black liberation in and through K-12 science education. *Journal of Science Teacher Education, 33*(2), 131–153. https://doi.org/10.1080/1046560X.2021.2008096
Mutegi, J. W. (2013). "Life's first need is for us to be realistic" and other reasons for examining the sociocultural construction of race in the science performance of African American students. *Journal of Research in Science Teaching, 50*(1), 82–103. https://doi.org/10.1002/tea.21065
National Academies of Sciences, Engineering, and Medicine. (2021). *Science and engineering in prekindergarten through fifth grades: The brilliance of children and the strengths of educators.* The National Academies Press. https://doi.org/10.17226/26215
Paris, D., & Alim, H. S. (2014). What are we seeking to sustain through culturally sustaining pedagogy? A loving critique forward. *Harvard Educational Review, 84*(1), 85–100.
Tate, W. (2001). Science education as a civil right: Urban schools and opportunity-to-learn considerations. *Journal of Research in Science Teaching, 38*(9), 1015–1028.

Chapter 5
Finding a Home in Science Education: Chronicle of My Crossroads Between the Zambian Sociocultural Perspectives and Formal Science Education

Vivien Mweene Chabalengula

5.1 Introduction

My journey to finding a home in science education has been guided by the profound words of Sandra Harding (1991): "What science? Whose science? What knowledge?" The "*What science?*" challenges the assumption that there is only one valid way of doing science, while "*Whose science?*" highlights the importance of considering the cultural and social context in which science is produced. "*Whose knowledge?*" emphasizes the need to recognize and value diverse forms of knowledge and experience. Harding's quote resonates deeply with me as my journey involved various crossroads where I had to acknowledge that formal science was viewed as a universal truth in science education settings, while at the same time figuring out how my own social cultural knowledge fit into the bigger science education picture. As a science student, educator and researcher of Zambian descent, I have often found myself navigating the complexities behind finding a real home and a sense of belonging in the scientific community. This quest for a "home" in science education has been a recurring theme throughout my journey, from my years as a high school science student to my current work as a science educator and researcher. As noted by researchers such as Bonilla-Silva (2006), the dominant Western epistemology in science education can often marginalize diverse perspectives and experiences, leading to a sense of displacement and disconnection for individuals from non-Western backgrounds. Yet, as argued by scholars like McKinley (2007), it is precisely this diversity of perspectives and experiences that can enrich our understanding of science and the world around us.

V. M. Chabalengula (✉)
University of Virginia, Charlottesville, VA, USA
e-mail: vmc4y@virginia.edu

M. M. Atwater and B. A. Garvin (eds.), *Creating and Maintaining a Home in Science Education Research*, Contemporary Trends and Issues in Science Education 65,
https://doi.org/10.1007/978-3-031-99382-4_5

In Zambia, where the rich cultural heritage of about 73 ethnic groups intersects with the legacy of colonialism and Western-dominated science education, the pursuit of scientific knowledge is often a complex and nuanced endeavor. While the Zambian sociocultural perspectives emphasize communal practices, values, spiritual connections to the land, and holistic understandings of the natural world (Mwanga, 2017), formal science education can be a foreign and exclusionary space (Hansen, 2018) because it emphasizes individualistic, empirical, and reductionist approaches (Aikenhead, 2001). Growing up, I was taught to revere the intricate relationships between humans, nature, and the spirit world. However, during my formal science education schooling, I struggled to reconcile my cultural heritage with the Western-dominated science education system because I was expected to embrace a scientific paradigm that often dismissed or marginalized these very beliefs. Yet, science education research has shown that one's cultural backgrounds and experiences influence their engagement and achievement in science education (Bayne, 2022; Hewson, 2015; Russell & Givens, 2022). To the contrary, science teaching practices often prioritizes Western knowledge systems, potentially marginalizing indigenous knowledge systems and cultural practices (Bang & Vossoughi, 2016; Harding, 2008). The dominance of Western epistemologies and methods can create a cultural mismatch for both educators and students from diverse backgrounds (Aikenhead, 2001; Harding, 2008). As a Zambian-born and diaspora science educator in the United States, I have navigated the crossroads of two worlds—the sociocultural heritage of my ancestors and the formal science education imparted through Western-style schooling. This crossroad raises important questions about the role of culture in science education, the need for inclusive and culturally responsive pedagogies, and the potential for affected individuals to reclaim and reframe scientific knowledge in ways that honor their own cultural identities and perspectives.

This auto-ethnographic chronicles my journey as I sought to reconcile these seemingly disparate worlds, finding a home in science education that honors both my cultural roots and the pursuit of scientific knowledge. Specifically, I discuss my Zambian sociocultural perspectives and formal science education, crossroads I encountered, how I navigated them, and how I found a home in science education. To contextualize my arguments, I drew on literature from science education, cultural studies, and social justice. Further, I illustrate how I found a home in science education via my personal anecdotes and experiences in my teaching and research activities. Finally, I discuss my experiences' implications to upcoming diaspora science educators and researchers by highlighting the need for a more inclusive and diverse approach to science education. I hope this chapter will spark critical discussions about the effects of the dominant science epistemology on the learning/teaching/research spaces among marginalized populations in science education.

5.1.1 *Zambian Sociocultural Perspectives and Formal Science Education*

The intersection between traditional Zambian sociocultural perspectives and formal science education creates a complex landscape not only in identity formation (Hansen, 2018), but also in teaching and learning practices. This cultural mismatch can lead to dissonance and marginalization, as educators and students alike navigate multiple worlds, reconciling cultural knowledge and practices with the demands of a science education system that often dismisses indigenous epistemologies (Collins, 1986; Harding, 2008; Nelson-Barber et al, 2022). In this section, I provide context for the readers so they have an idea of Zambian social cultural perspectives and how it intersects or parallels formal science education. Three major areas that embody my cultural heritage that tend to clash with formal science education are cultural practices, beliefs about natural phenomena, and values.

5.1.2 *Cultural Practices in Nature Conservation, Farming, Medicine*

Zambian cultural practices, particularly, in nature conservation, farming, and medicine, play a significant role in the well-being of the people and the natural environment. For instance, in nature conservation, local communities have traditional knowledge of conservation and management of natural resources, such as protecting sacred forests (shrines) and wildlife. This practice aligns with scientific concepts in ecology and conservation biology (Mwanga, 2017). However, the disconnect is in the approaches to conservation. Traditional Zambian cultural practices may prioritize spiritual and cultural values over scientific conservation methods, and indigenous knowledge may not be recognized or valued in formal science education and conservation efforts (Kassam, 2010), even if it can inform scientific research and management of natural resources. Traditional farming practices such as crop rotation and organic farming are used to promote soil fertility and conserve water. These practices align with scientific concepts in agriculture and soil science. However, traditional Zambian farming practices may not be compatible with modern scientific farming methods, leading to conflicts over land use and resource management (Hikwa, 2018). As such, indigenous knowledge may be marginalized or replaced by scientific farming practices, leading to loss of cultural heritage (Kassam, 2010). Additionally, traditional medicine practices use local plants and herbs to treat various ailments and have been found to be effective in treating certain diseases. The use of local plants and herbs align with scientific concepts in botany, pharmacology, and medicine. However, traditional medicines are not recognized or valued in formal science education and healthcare systems (Chinsembu, 2016; Kassam, 2010).

5.1.3 Cultural Beliefs About Natural Phenomena

In my culture, there are distinct cultural beliefs, particularly those surrounding natural phenomena such as disease causation and the sun and moon. Specifically, the sun and moon are associated with ancestral spirits, deities, or cultural heroes, and their movements and phases are believed to influence daily life, agriculture, and rituals. As expected, my beliefs were in stark contrast to the Western worldview where the sun and moon are perceived as celestial bodies governed by physical laws, with their movements and phases explained by astronomy and physics. To provide context, in traditional Zambia, disease causation is attributed to the ancestral spirits or curses, which conflict with scientific concepts of microbiology and epidemiology (Chinsembu, 2016). Similarly, indigenous beliefs and knowledge about celestial bodies (such as the sun, moon) and natural phenomena may be seen as unscientific or primitive (Jegede & Okebukola, 1991). Yet, these are important beliefs that connect an individual to his/her culture. However, formal science education does not integrate nor make references to these perspectives and are not seen as a valid way of understanding the natural world among the local people (Kassam, 2010). An empirical study by Jegede and Okebukola (1991) showed just how important cultural beliefs can impact one's science learning. Jegede et al. investigated 319 science students from West African universities using the Traditional Cosmology Test (TCT) and the Test of Observational Skills (TOS) to determine whether science observation skills can be influenced by students' belief in traditional African cosmology, beliefs and superstitions. The results showed that students who had strong beliefs in African traditional cosmology made significantly fewer correct observations compared to those with a low belief level. These authors postulated that traditional cosmology could have been a major factor responsible for this performance difference in observation. Jegede et al. study brings to the fore the importance and centrality of learners' cultural values in the teaching/learning process if formal science knowledge is to be assimilated and accommodated without causing much dissonance and confusion. After all, science is also just another way of knowing the world! Reflecting on my own learning process, my high school and university science teachers who happened to be Zambians, did not present the material from both viewpoints. This disconnect between my cultural knowledge and western science has been the pivotal rationale for my research interests in multicultural science education.

5.1.4 Cultural Values

My culture values community and collective knowledge-sharing contrasted with the individualistic approach often emphasized in Western science education. My cultural background emphasizes the importance of community and collective success. As a science student, educator and researcher, I was encouraged to collaborate with peers and seek help from teachers and mentors. This communal approach to learning coupled with my cultural heritage values on knowledge-sharing has shaped me to be a constructivist science educator. However, several researchers (e.g., Avraamidou, 2020; Bang & Medin, 2010) acknowledge that science education has overlooked the critical role of social cultural processes that shape students' science identities, experiences and outcomes, learning strategies, and epistemological commitment. For example, in his conceptual paper, Avraamidou (2020) argued that science identity is a landscape that emphasizes recognition of one's values and emotions, both of which can give meaning to the process of becoming a science person. According to Avraamidou, value recognition refers to how an individual is viewed by others within a given sociopolitical context including specific cultural norms, values, beliefs, and stereotypes. This intertwined web, which I would refer to as a "social-scientific political web" (SSPW) plays a major role in shaping one's science identity. My argument is that this SSPW gives rise to Harding's (1991) still unanswered questions: "What science? Whose Science? Whose Knowledge?" Additionally, the SSPW raises a critical question relevant to this chapter: "Who is allowed to enter and stay in the science education home"? This later question is directly linked to Avraamidou's (2020) *emotion* aspect of the science landscape. Emotions can build or break one's chances of fitting in or not in the science education home depending on how one responds to the systemic inequalities of the SSPW (such as racism and gender bias) because "forming a science identity is not only personal, but also political" (Avraamidou, 2020). In my own experiences as a Zambian and diaspora science student, educator and researcher, understanding the dynamics of the SSPW has helped me understand how science is intertwined with social, political, and cultural contexts. It has taught me to approach science as a complex value system and multifaceted endeavor that requires collaboration, adaptability, and a commitment to social responsibility.

5.2 Integration of Social Cultural Perspectives in Science Curriculum

Though some researchers have found that traditional Zambian cultural practices and knowledge systems have significant impact on students' understanding and engagement with science education (Mwanga, 2017), my formal science education did not integrate it. As a student, I felt like an "alien" to science because of the way it was taught and the extent to which my social cultural perspectives were integrated in the teaching and learning process. My experiences of feeling like an outsider to science learning are summed up well in Banks' (2016) four approaches for the integration of multicultural content into the curriculum (*contributions*, *additive*, *transformation*, and *social action* approaches). Particularly, the contributions and additive integration approaches resonate well with me. In the *contributions* approach, multicultural content is limited to discussing ethnic heroes or holidays from various cultures so much that the content is simply an appendage to the mainstream curriculum without changing the curriculum content. In the *additive* approach, multicultural content is added to the curriculum, but taught/learned from the mainstream scientific viewpoint. From a social cultural standpoint, both the contributions and additive approaches can limit students' understanding of science from different perspectives because the "dominant" effect comes into play. That is, the mainstream context tends to diminish the "other" perspectives and skews how people view science and the world. To the contrary, the *transformative* approach challenges the dominant structure of the science curriculum and empowers both teachers and students to understand science concepts and themes from different cultural perspectives. In science education settings, this implies that learners and teachers have the opportunity to be aware and understand that science is just one way of knowing among others such as cultural or indigenous knowledge. With respect to the *social action* approach, it not only includes all the elements of the transformative approach, but also adds components that require teachers and students to make decisions and take actions related to their understanding of natural phenomena. In this approach, science curriculum would provide opportunities for students to utilize knowledge sources from both their social cultural and formal science knowledge bases to brainstorm, decide, and make personal/societal decisions and to take actions related to the content, concepts, issues, and problems they are investigating—thereby leading to social justice science curriculum, which is about relevance, rigor and empowerment (Atwater, 2010).

Reflecting on my science learning experiences, I never had a chance to learn science where I actively and meaningfully engaged with my cultural beliefs and practices. The Zambian science curriculum was modeled after the British perspectives (as Zambia was a British colony until October 1964). My science education did not prepare me to think through these concepts from my own ecological background to be empowered and take ownership of my own learning. Bang and Vossoughi (2016) also argued that if Zambian cultural practices were meaningfully integrated into formal science education, educators and students could develop a deeper understanding of the natural world and the interconnectedness of traditional and scientific knowledge systems. Further, meaningful integration of sociocultural perspectives have the potential to promote inclusivity, equity, and a sustainable and culturally responsive science education system.

Out of curiosity and interest in multicultural education in recent years as a researcher, I set out to investigate whether the Zambian science curriculum is now integrating the local sociocultural perspectives where learners could relate and see the relevance to their own immediate environment within science education themes. I examined the integration of multicultural science content and context in Zambia's high school biology curriculum materials (Chabalengula & Mumba, 2022) using Banks' integration approaches. The results revealed that Zambian sociocultural perspectives were integrated at the additive and contributions levels, and not at the transformation or social action levels in Teachers' Guidebooks. Even though many science concepts in the Zambian science curriculum (notably, nutrition, health and disease, conservation of forests and ecosystem, genetics and inheritance) pose a crossroad for both educators and learners in terms of where their sociocultural perspectives fit in, the curriculum guidelines do not promote the engagement with both the scientific and traditional belief perspectives. Furthermore, learners are rarely engaged in crafting feasible actions applicable to the Zambian contexts. A good example was evident in the topic "deforestation". Simply telling a typical Zambian student from a rural area that cutting down trees is not good would be a disservice to their way of life, since rural people use timber for their basic needs and survival (e.g., firewood, utensils, house furniture). Rather, learners need to be engaged in a social action approach where they would weigh options and craft a feasible action plan toward combating deforestation, but within the undisputable reality that people still need to utilize trees/timber for survival (social action approach).

5.3 Navigating the Crossroad

5.3.1 Why I Initially Felt "Homeless" in Science Education

Due to the way I learned formal school science (notably memorization of facts without real-life applications or meaningful integration of my cultural knowledge), I was compelled to believe that cultural/indigenous knowledge and western science are two different worlds and oceans apart. In reality, these two worlds are closely interrelated because science is not just a product of individual scientists and their discoveries, but shaped by the culture in which it is practiced. Further, both science and cultural perspectives are simply some of the different ways of knowing and understanding our natural world. Interestingly, science and cultural knowledge share several fundamental attributes—that is both are constantly verified through repetition and verification, inference and prediction, empirical observations, and recognition of patterns for natural phenomena/events.

Regrettably, science education literature shows that there is a conflict between cultural knowledge and the Western scientific paradigm that is often expressed through "traumatic and destructive hybridity" (Langer, 2011), which refers to the harmful and violent effects of cultural hybridity, where the blending of cultures results in the erasure, suppression, or destruction of indigenous cultures and identities. In the context of science education, traumatic and destructive hybridity can manifest in several ways: cultural suppression (indigenous knowledge and practices are marginalized or replaced by Western scientific approaches, leading to cultural erasure and loss of identity); cultural appropriation (Western science appropriates and commodifies indigenous knowledge without proper understanding or respect perpetuating cultural exploitation); and epistemic violence (Western scientific knowledge is imposed as the sole legitimate perspective, dismissing and devaluing indigenous epistemologies and ways of knowing). In my case, cultural suppression has been the most evident since my student days because the teaching/learning approaches and science curriculum materials did not integrate my culture at higher levels, and the practice is still existing as evidenced in a recent study of Zambian science curriculum (Chabalengula & Mumba, 2022).

In my recent roles as a science educator and researcher, cultural appropriation and epistemic violence have had an impact on myself, especially at the beginning of my career in the United States. Many times, I felt like my teaching philosophy (constructivist, experiential and participatory) and research focus (enhancing scientific literacy and integrating multicultural science content/contexts) were not aligned to the mainstream teaching and research viewpoints. The fact that my cultural background has had tremendous influence on my personal-self made me feel like an imposter, much like what Clancey (2019) defined as imposter syndrome. Despite several achievements, I would sometimes downplay and doubt my abilities, revisit my cultural displacement, and believe it or not still re-assessing my English language fluency as English is my fourth language! However, my growth mindset and willingness to "feel" these emotions have provided opportunities for me to be receptive to cultural exchange and adaptability, thereby making it easier for me to find a home in the science education community.

5.3.2 *Identity Formation in Science Education*

My identity in science education started forming during my graduate school years in the United States. My turning point came when I took curriculum theory courses that exposed me to Freire's (1970) critical pedagogy and Ladson-Billings' (1995) culturally relevant teaching, as a new graduate student. At that point, I realized that science education could be a tool for social justice and empowerment, rather than merely a transmission of Western knowledge systems. Freire's critical pedagogy has been instrumental in my identity formation as a science educator and researcher. His approach emphasizes the importance of critical reflection, dialogue, and problem-posing, which has helped me to interrogate the dominant Western scientific paradigm and reclaim my cultural identity. Through critical pedagogy, I began to recognize the cultural biases and power dynamics that shape science education and have learned to challenge and disrupt these oppressive systems. By integrating my cultural knowledge and experiences into my teaching and research practices, I found a sense of empowerment and belonging in science education.

Ladson-Billings' (1995) concept of culturally relevant teaching has been instrumental in my identity formation as a science educator and researcher. Her approach emphasizes the importance of understanding and valuing students' diverse cultural backgrounds, experiences, and knowledge systems (Ladson-Billings, 1995). By incorporating culturally relevant teaching into my teaching practice, I have created an inclusive learning environment that acknowledges and respects my students' cultural identities (Gay, 2000). This approach has helped me to reconcile my own cultural heritage with my role as a science educator, and to find a sense of belonging and purpose in science education (Hord, 2017). Culturally relevant teaching has also enabled me to challenge the dominant Western scientific paradigm and promote a more inclusive and equitable science education that values diverse ways of knowing and being (Bang & Vossoughi, 2016).

5.3.3 *Finding a Home in Science Education*

Critical pedagogy and culturally relevant teaching enabled me to form a strong identity as a science educator and researcher, one that is rooted in my cultural heritage and committed to promoting social justice and inclusivity in science education. Through critical reflection and self-awareness, I have come to embrace my diverse identity and found a sense of belonging in science education by integrating my cultural knowledge and experiences into my teaching and research practices. This journey has allowed me to find a home in science education, where I can authentically express myself and contribute to the development of inclusive and culturally relevant science education. As Davies (2008) noted, having a strong sense of identity and purpose helped me navigate the complexities of being a diaspora science educator and researcher and enabled me to make a positive impact on my students and the wider science education community. I found my foot and home in science education by embracing my socio-cultural heritage and integrated it into my work—teaching and research. This integration has created a seamless flow of who I am so much that I do not see a "gap" in my identity and sense of belonging. I no longer feel like an "imposter" because I have learned to mesh my true self and the work I do. Specifically, I now teach science education courses from a lens of a pedagogy of belonging, where I aim to create learning spaces for my students to feel a sense of belonging, safety, and connection, fostering a sense of community and shared purpose. In terms of research, I intentionally investigated themes that draw from my cultural experiences. For instance, my research interests in scientific literacy and multicultural science education enable me to immerse myself in what I cared about while also fitting in within the broader science education community. One takeaway message I have learned from my experiences is that finding a home in science education as a diaspora science educator and researcher requires self-awareness, empowerment, and a willingness to challenge the status quo. By embracing my cultural heritage and indigenous knowledge systems, I found a sense of belonging and purpose in science education. Below are some examples of things I did to try and fit in the science education community.

5.3.4 *As a Graduate Science Education Student*

As highlighted earlier, reading and examining Freire's (1970) Critical Pedagogy exposed me to a journey of self-discovery and empowerment. During graduate school, I intentionally began to explore Zambian indigenous knowledge systems and cultural practices, integrating them into my course assignments. For my doctoral dissertation, I explored the broad topic themes: scientific literacy and science concept understanding, but from the lens of the Zambian context. This approach ignited in me a sense of self-awareness and gave me an opportunity to really examine how socio-cultural perspectives influence science education in my social-cultural context. By embracing my cultural heritage and indigenous knowledge systems, I found a sense of belonging and purpose in science education. Specifically, I examined the nature and extent of scientific literacy (SL) themes coverage in Zambian national high school biology curriculum (biology textbooks, biology syllabi, and grade twelve national biology examination papers) (Chabalengula et al., 2008). The results showed that the biology textbooks and syllabi emphasized basic knowledge of science while the biology examination papers emphasized science as a way of knowing. The interaction between science, technology and society theme was the least represented in the Zambian biology curriculum, which I argued could have been a barrier to a better preparation of scientifically literate citizens in the digital-driven world. This study immersed me in not only examining my own learning experiences, but also helped me learn that SL themes were not well represented in other education systems worldwide. As such, I published this research in an international journal because it was relevant to both my social-cultural context and to the larger science education community. This was the beginning of my confidence that I could fit in in science education. At this juncture, I need to stress the importance of mentorship. During my graduate studies, I was fortunate to have faculty mentors and fellow graduate students who not only supported me but also respected and valued my social-cultural perspectives on the learning/teaching issues in science education. This support network heightened my sense of belonging and connection to a community of science education learners and scholars. Russell and Givens (2022) also highlighted the role of mentoring, sense of belonging, and self-efficacy in promoting persistence and resilience for African (American) graduate students.

5.3.5 As a Diasporic Science Educator

Early on as a science educator, I discovered the work of pedagogy researchers like Ladson-Billings (1995), who emphasized the importance of critical pedagogy and culturally relevant teaching. Through my teaching, I have sought to integrate my acquired knowledge in critical pedagogy and my cultural heritage in my science education practice, creating a sense of home and belonging in the field. I have drawn on knowledge about how diverse people learn to develop innovative and relevant course syllabi and teaching strategies. I embrace a critical and reflexive approach to science education. By recognizing and valuing diverse sociocultural perspectives, I aimed to create inclusive learning environments that honor students' cultural identities and their local environments.

Over the years, my teaching philosophy has been informed by three avenues. First, my lived experiences and interactions with diverse peoples have made me realize that every learner, regardless of cultural background, has the capability to learn, and that each student learns in different ways and pace. I gradually realized that for equitable learning to occur, it is my duty to provide a positive and inclusive learning space that embraces learners' diverse perspectives for them to be actively engaged in the construction and application of knowledge. Second, my scholarship agenda on enhancing scientific literacy, inquiry-based instruction approaches, and integrating multicultural science content and contexts in science education, has enlightened and exposed me to various research-based understandings about how diverse people experience and navigate the learning process. With this research knowledge base, I have come to understand that effective teaching approaches are those that provide real-life, sense-making learning contexts to promote meaningful experiential learning. More recently, my experience teaching fully online due to COVID-19 pandemic has made me rethink how the course materials and content delivery can be enhanced to make the learning environment more participatory online. In sum, I describe my teaching philosophy as Constructivist, Experiential and Participatory (CEP). To implement my CEP teaching approach in a coherent and organized manner, I incorporate the inquiry-based 5E Learning Cycle Model (Engagement, Exploration, Explanation, Extension, and Evaluation) across all my courses (Bybee, 2015). In relation to my CEP teaching philosophy, the 5E learning cycle provides avenues for students to construct knowledge, have direct experience with the content, and be active participants in the learning process. The learning cycle also provides a differentiated-learning environment in that learners can demonstrate their understandings in different modalities via the 5E phases. Together, my CEP teaching philosophy and the 5E learning model helps me attend to diverse learning needs and preferences, and consequently enable my students to be fully engaged and take ownership of their own learning, while at the same time being challenged to interact and apply course content to their immediate surroundings. My CEP approach seems to align with Nelson-Barber et al. (2022) concept of place-based learning, which emphasizes the importance of grounding education in the local context, culture, and environment. Nelson-Barber et al.'s place-based learning approach integrates Indigenous and community knowledge to promote cultural responsiveness and inclusivity, fosters a sense of belonging among students, encourages hands-on and experiential learning of local issues that enables students to become active contributors to their communities.

To illustrate how I implement my CEP teaching approach, I describe a typical module progression in one of my science education graduate courses I teach titled "Trends and Issues in Science Education". This course is asynchronous with students residing in various parts of the world including the United States, Japan and China. Specific diversities in this course are linguistic, cultural, and science educational systems—all of which required me to critically review my materials often to make sure they are relevant and applicable to each learner, regardless of their educational and cultural ecology. On a positive note, this diversity makes class discussions informative and intriguing. In the module "Science Education (ScEd) Curriculum Frameworks", I start by engaging students to explore the ScEd curriculum framework(s) in their respective countries via a pre-module activity. This exercise allows them to not only know more about their local curriculum, but also have a sense that their own knowledge and contributions are highly valued in the course. As such, this activity aligns well with the engagement and exploration phases of the 5E learning model. Thereafter, I provide opportunities for students to demonstrate their understanding of their local curricula frameworks in discussion posts and assignments. These reading materials and personalized assignments enable students to be engaged, construct knowledge, see relevance in the course content, and get to learn more about other non-USA frameworks from peers. Throughout the semester, students complete different assignments, individually and collaboratively. One of the criteria I integrated to make learning meaningful, relevant, and applicable was to provide options for students to complete the assignments from the lens of their actual workplace settings and/or local ScEd policies or practices. Allowing students to complete assignments using their own local environments/contexts provided opportunities for them to take a step back and make meaningful connections to the course content.

In relation to finding a home in science education, having an opportunity to teach and work with people from diverse groups enables me to have this natural connection to my own socio-cultural experiences in science education. My teaching approach not only diminishes dissonance and marginalization among students, but also they empower themselves to take ownership of their own learning. I also believe that my approach attempts to let students have an open mind to reflect on Harding's (1991) quote: "*What science? Whose science? Whose knowledge?*" I argue that creating this reflective space is critical if learners are to critically examine how and where and their sociocultural perspectives fit into formal science education.

5.3.6 As a Diasporic Science Education Researcher

As discussed earlier, my doctoral work in graduate school helped me establish my research agenda and gave me an edge to start mapping a home in science education. However, I have expanded my research strands to include topics and themes that not only aligns with my sociocultural context as a diasporic science education researcher, but also relevant to the science education community. Doing so has enabled me to leverage my unique perspective and experiences, and provided a space or home for me to contribute to the diversification of research insights and thought-perspectives in science education research (Hutchinson, 2002).

Early on in my research career, I investigated students' understandings of science concepts that tend to pose learning difficulties, particularly the "energy concept" (Chabalengula et al., 2012b) and "biotechnology" (Chabalengula et al., 2011). The insights I gained from these early studies were instrumental in helping me study how various misconceptions on science concepts could be addressed using inquiry-based instruction approaches—leading to my research activities in inquiry science teaching during my mid-career years as a diasporic science education researcher. Within the inquiry science teaching research endeavors, I started out by examining coverage of inquiry skills in Zambian science curriculum materials (Chabalengula & Mumba, 2012a), and American science teachers' conceptual understandings and performance on inquiry skills (Chabalengula et al., 2012a). These studies revealed to me that teaching science by inquiry as a research topic cuts across geographic regions—in this case from both the Zambian and American context. These studies positioned me well and enabled me to begin "settling down" in the science education community. Later, my focus became finer as I started investigating specific inquiry approaches, namely model-based inquiry instruction (Chabalengula & Mumba, 2012b), and computer simulations (Chabalengula et al., 2016). The release of the Next Generation Science Standards (NGSS) in the United States (NRC, 2012) created an exciting opportunity for me to examine inquiry-based science instruction from the lens of science and engineering practices, and engineering design. To start, I wrote two internal grant proposals with internal funding from the School of Education Research and Development Fund in academic years 2014–2015 and 2017–2018 at the University of Virginia. These projects were significant and timely because there was limited literature on the new NGSS. Through these grant projects, I was able to unpack the nature and extent to which the new science and engineering practices and design skills are addressed in widely used K-12 engineering programs in the United States such as Engineering is Elementary and Learning by Design (Chabalengula & Mumba 2017; Chabalengula et al., 2017). Taken together, the findings from my studies on inquiry-based science instruction have been the backdrop through which I now utilize the 5E learning cycle to organize my teaching and utilize various tasks to provide diverse opportunities for students to interact with course material and construct knowledge for themselves.

Drawing on my own past experiences where my social cultural perspectives were absent from the formal science curriculum, I have been reading a lot about multicultural science education literature in recent years. As I discussed earlier in the chapter, Banks' (2016) four approaches for the integration of multicultural content into the curriculum (i.e., *contributions*, *additive*, *transformation*, and *social action*) were invaluable to me. This is because I was compelled to reflect on my science education journey as a high school student, university student, science educator and researcher. I was also taken back to Harding's (1991) questions: "*What science? Whose science? What knowledge?*" As such, my scholarly work has expanded to include investigations of *how* and *when* multicultural science content/contexts are integrated in science curriculum materials—again starting with the Zambian science curriculum (Chabalengula & Mumba, 2022). The results I gleaned from this study were significant as they informed me, as a science educator and curriculum designer, that it is not enough to simply use multicultural content/contexts as "add-ons"; instead, meaningful integration must occur if students are to utilize their rich cultural ecology as a sense-making learning context to construct knowledge, encounter experiential learning, and participate fully in the learning process. Through this research, I found a sense of purpose in science education, recognizing the critical role that multicultural science content and contexts play in shaping students' identities, interests, and career aspirations (Atwater, 2022).

Reflecting back on my research pathways, I realize that pursuing interests that matter to me while also being relevant to the larger science education community have been pivotal in my finding a home in science education. As Bayne (2022) alluded, the intersection of personal and professional identities and moments of inspiration and motivation can play a critical role in promoting the global representation of underrepresented scientists of color and their successes, and consequently enable them to establish a home in the science community. To date, I have several peer-reviewed journal articles, book chapters in edited volumes, and conference presentations at international and national meetings for several organizations including the National Association for Research in Science Teaching (NARST), and Association for Science Teacher Education (ASTE). I have been a principal investigator or co-investigator on several internal and external grant funding agencies including National Science Foundation (NSF), State Council of Higher Education for Virginia (SCHEV), Illinois State Board of Education (ISBE), and UVA's School of Education and Human Development Dean's grants. All these dissemination outlets cemented my home in science education.

5.3.7 *Implications to Upcoming Science Education Educators and Researchers*

My journey in finding a home in science education has been both a crossroad and a winding path. However, my research themes on scientific literacy, inquiry teaching approaches and multicultural science education have been a beacon of light that has helped me find my way and be established in science education. This is because the insights I gained from my research activities not only informed me about how diverse learners learn science, but also empowered me to design science lessons that are culturally relevant and create teacher-student relationships that are respectful and welcoming. As such, one of the most important lessons I have learned in my journey is ensuring that my academic roles and responsibilities align and inform each other to create a seamless transition among them. Doing so not only enabled me to demonstrate how well I am prepared as a science educator and researcher. Further, this has served me well in my job where I am evaluated for my effectiveness in teaching, research and service every year.

My teaching approaches that embrace different students' input help me empower students to take charge of their own learning, thereby paving the way for more diverse and inclusive perspectives in science education and research. My research interests highlight the value of authenticity and finding one's own place in science education, encouraging others to do the same. My professional identity as both a Zambian and diasporic science educator and researcher underscores the importance of considering intersectional identities and experiences in science education and research. Above all, my chronicle demonstrates the importance of resilience and determination in establishing a home in science education.

My experiences as a Zambian and diasporic science educator and researcher can serve as a powerful inspiration, empowering the next generation of science educators and researchers underrepresented in the sciences to find their own place and voice in the scientific community. Notable aspects evident in my chronicle include representation, cultural connection, contextualized learning, authenticity, and intersectionality. As one who belongs to the underrepresented groups in the sciences, I believe this chronicle of my social cultural experiences and perspectives helps bring to the fore a myriad of crossroads between one's cultural background and scientific knowledge, and how one can navigate those crossroads. My lived experiences and educational background in different educational systems provides me with an understanding of how diverse educational systems and cultural contexts can enrich science education, making it more inclusive and effective for students from marginalized or oppressed groups.

5.4 Conclusion

My journey as a diasporic science educator and researcher has been marked by a quest to find a sense of home and belonging in the field of science education. Through critical reflection and examination of my experiences, I have come to realize that finding a home in science education requires navigating the crossroad between one's cultural heritage and challenging dominant Western epistemologies. As a Zambian science educator and researcher, I embraced critical reflection, cultural responsiveness, and a commitment to challenge the status quo by following research interests that matter to one's socio-cultural realm, while also being relevant topics to the science education community. I urge fellow science educators to recognize and value diverse sociocultural perspectives, promoting inclusive science education. As a diaspora scientist, I have had to navigate the crossroad between my Zambian cultural heritage and the Western-dominated science education system. Through my research interests and teaching approaches, I have sought to integrate my cultural knowledge and experiences into my science education practice, creating a sense of home and belonging in the field.

My journey underscores the need for science education educators and researchers to prioritize their cultural knowledge and experiences. Finding a home in science education as a diasporic Zambian science educator and researcher required me to have critical reflection, cultural responsiveness, and a commitment to relate what matters to me as well as to the science education community towards a common goal of promoting inclusivity and equity. By embracing our diverse cultural knowledge and experiences as a people, we can work towards creating a more just and equitable science education system that values and empowers all regardless of cultural and geographical diversities.

References

Aikenhead, G. S. (2001). Integrating Western and indigenous knowledge systems: Implications for science education. *International Journal of Science Education, 23*(5), 577–591.

Atwater, M. M. (2010). Multicultural science education and curriculum materials. *Science Activities, 47*, 103–108.

Atwater, M. M. (Ed.). (2022). *International handbook of research on multicultural education* (Vols. 1 and 2). Springer.

Avraamidou, L. (2020). Science identity as a landscape of becoming: Rethinking recognition and emotions through an intersectionality lens. *Cultural Studies of Science Education, 15*, 323–345. https://link.springer.com/article/10.1007/s11422-019-09954-7

Bang, M., & Medin, D. (2010). Cultural processes in science education. *Journal of Research in Science Teaching, 47*(6), 673–699.

Bang, M., & Vossoughi, S. (2016). Participatory science and the politics of empowerment. *Science Education, 100*(6), 1091–1111.

Banks, J. A. (2016). Approaches to multicultural curriculum reform. In J. A. Banks & C. A. Banks (Eds.), *Multicultural education: Issues and perspectives* (9th ed., pp. 151–168). Wiley.

Bayne, G. U. (2022). Global voices: Personal and professional lived experiences of Black and Brown women's culture in science. In M. M. Atwater (Ed.), *International handbook of research on multicultural education* (pp. 1447–1482). Springer. https://doi.org/10.1007/978-3-030-83122-6_45

Bonilla-Silva, E. (2006). *Racism without racists: Color-blind racism and the persistence of racial inequality in the United States*. Rowman & Littlefield.

Bybee, R. W. (2015). *The BSCS 5E instructional model: Creating teachable moments*. NSTA Press.

Chabalengula, V. M., Bendjemil, S. A., Mumba, F. & Chiu, J. L. (2017). Nature and extent of science and engineering practices coverage in K-12 engineering curriculum materials. *International Journal of Engineering Education, 33*(1B), 308–320. Special Issue on Current Trends in K-12 Engineering Education.

Chabalengula, V. M., Fateen, R., Mumba, F., & Ochs, L. K. (2016). Effect of inquiry-based computer simulation modeling on pre-service teachers' understanding of homeostasis and their perceptions of design features. *Journal of Computers in Mathematics and Science Teaching, 35*(3), 225–248.

Chabalengula, V. M., & Mumba, F. (2012a). Inquiry-based science education: A scenario on Zambia's high school science curriculum. *Science Education International, 23*(4), 307–327. Special Issue on Inquiry Based Science Education. https://files.eric.ed.gov/fulltext/EJ1001626.pdf

Chabalengula, V. M. & Mumba, F. (2012b). Promoting biological knowledge generation using model-based inquiry instruction. *International Journal of Biology Education, 2*(1), Article 2.

Chabalengula, V. M., & Mumba, F. (2017). Engineering design skills coverage in K-12 engineering program curriculum materials in the USA. *International Journal of Science Education, 39*(16), 2209–2225. https://doi.org/10.1080/09500693.2017.1367862.6

Chabalengula, V. M., & Mumba, F. (2022). Multicultural science content and contexts in Zambian science curriculum materials. In M. M. Atwater (Ed.,), *International handbook of research on multicultural science education* (Chapter 28). Springer. ISSN 2197-1951. https://doi.org/10.1007/978-3-030-37743-4_28-1

Chabalengula V. M., Mumba, F., & Chitiyo, J. (2011). Elementary education pre-service teachers' understanding of biotechnology and its related processes. *Biochemistry and Molecular Biology Education, 39*(4), 321–325. https://eric.ed.gov/?id=EJ959423

Chabalengula, V. M., Mumba, F., Lorsbach, A., & Moore, C. (2008). Curriculum and instructional validity of the scientific literacy themes covered in Zambian high school biology curriculum. *International Journal of Environmental and Science Education, 3*(4), 207–220. https://files.eric.ed.gov/fulltext/EJ894861.pdf

Chabalengula, V. M., Mumba, F., & Mbewe, S. (2012a). How pre-service teachers understand and perform science process skills. *Eurasia Journal of Mathematics, Science and Technology Education, 8*(3), 167–176.

Chabalengula, V. M., Sanders, M., & Mumba, F. (2012b). Diagnosing students' understanding of energy and its related concepts in biological context. *International Journal of Science and Mathematics Education, 10*(2), 241–266. https://link.springer.com/article/10.1007/s10763-011-9291-2

Chinsembu, K. C. (2016). Traditional medicine in Zambia: An ethnobotanical survey. *Journal of Ethnobiology and Ethnomedicine, 12*(1), 1–15.

Clancey, S. (2019). The imposter phenomenon in higher education: A review of the literature. *Journal of Further and Higher Education, 43*(1), 34–45.

Collins, P. H. (1986). Learning from the outsider within: The sociological significance of Black feminist thought. *Social Problems, 33*(6), 514–532.

Davies, B. (2008). Subjectification and the pedagogical relation. In B. Davies & S. Gannon (Eds.), *Pedagogical encounters* (pp. 15–34). Peter Lang.

Freire, P. (1970). *Pedagogy of the oppressed*. Herder and Herder.

Gay, G. (2000). *Culturally responsive teaching: Theory, research, and practice*. Teachers College Press.

Hansen, T. (2018). Science education and cultural diversity: A study of Zambian students' experiences. *Journal of Research in Science Teaching, 55*(6), 831–854.

Harding, S. (1991). *Whose science? Whose knowledge? Thinking from women's lives*. Duke University Press.

Harding, S. (2008). *Sciences from below: Feminisms, postcolonialism, and modernities*. Duke University Press.

Hewson, M. (2015). Cultural diversity and science education: A review of the literature. *Studies in Science Education, 51*(1), 1–35.

Hikwa, L. (2018). Indigenous knowledge and sustainable agriculture in Zambia. *Journal of Agricultural Education and Extension, 24*(3), 257–273.

Hord, S. M. (2017). African American women in STEM fields: Experiences and perceptions. *Journal of Women and Minorities in Science and Engineering, 23*(2), 147–162.

Hutchinson, S. (2002). The importance of cultural context in science education. *Science Education, 86*(5), 642–655.

Jegede, O. J., & Okebukola, P. A. (1991). The relationship between African traditional cosmology and students' acquisition of a science process skill. *International Journal of Science Education, 13*(1), 37–47. https://doi.org/10.1080/0950069910130104

Kassam, K. (2010). Indigenous knowledge and development: A postcolonial critique. *Journal of Postcolonial Studies, 13*(2), 135–154.

Ladson-Billings, G. (1995). But that's just good teaching! The case for culturally relevant pedagogy. *Theory into Practice, 34*(3), 159–165.

Langer, J. (2011). Indigenous knowledge and western science. In *Postcolonialism*. Palgrave Macmillan. https://doi.org/10.1057/9780230356054_6

McKinley, E. (2007). Postcolonialism, indigenous students, and science education. In S. K. Abell & N. G. Lederman (Eds.), *Handbook of research on science education* (pp. 199–226). Lawrence Erlbaum Associates.

Mwanga, N. (2017). Traditional ecological knowledge and sustainable development in Zambia. *Journal of Environmental Studies and Sciences, 7*(2), 147–157.

Nelson-Barber, S., Johnson, C., Boxerman, J., Silberglitt, M., & Rechebei, E. (2022). Using context-adaptive indigenous methodologies to address pedagogical challenges in multicultural education. In M. M. Atwater (Ed.), *International handbook of research on multicultural education* (pp. 711–738). Springer.

NRC (National Research Council). (2012). *A framework for K-12 science education: Practices, crosscutting concepts, and core ideas*. The National Academies Press.

Russell, M., & Givens, M. (2022). Factors that impact persistence and the culture of higher education. In M. M. Atwater (Ed.), *International handbook of research on multicultural education* (pp. 1405–1424). Springer.

Chapter 6
A Home of Steel and Velvet: An African Professor's Insights for the Successful Navigation of the U.S. Academic Landscape

Charles B. Hutchison

6.1 Genesis

Being an African who hails from a tribe known for her pleasantries and gentility of spirit, it has been enlightening to discover of myself that I have a competitive spirit that is ensconced in an amalgam of steel and velvet (cf. Andelin, 1994): steel, because it is stable, strong, and resilient, and velvet, because it is gentle to the touch and offers comfort to its hosts. If you had asked the steel to describe itself, it could not, because it had not yet been tested—but we were yet to find out.

My journey to the place I ultimately called my academic home began as an unconscious process that I suspect is shared by many in the profession. The road to this destination, however, was strewn with seeds that only became visible when my thoughts were stimulated by reflective promptings. An inspection of these seeds pronounced the inevitability that I had to become a professor, and to never rest until I had attained the highest credentials in the profession, but in the most parsimonious, optimized, and measured way. Along the way, I learned that the best of them are often not too far from most of them, and that an extra five percent of the proper ingredients can help most people to become the best of them. This is the story I am still learning and is still unraveling, nested in steel, and clad in velvet.

C. B. Hutchison (✉)
The University of North Carolina at Charlotte, Charlotte, NC, USA
e-mail: chutchis@charlotte.edu

M. M. Atwater and B. A. Garvin (eds.), *Creating and Maintaining a Home in Science Education Research*, Contemporary Trends and Issues in Science Education 65,
https://doi.org/10.1007/978-3-031-99382-4_6

6.2 Origins and Context

It is said that there are many ways to skin a cat, and so it is with personal narratives. The trajectories of one's life can be followed along numerous axes—so much so that it is difficult to delineate the best path for one's life narrative. Fortunately, this volume solicits one's academic and research life narrative, and so it will be.

The prompted retrospection of my academic and research life compelled me to uncover that my life's journey from being a science student to becoming a science educator has rested in the delicate embrace of time: both in the academic and non-academic spheres, my journey has been largely characterized by fragile calculi of average challenges, narrow escapes, and above-average fortunes, by all reasonable and logical assessments. There were many times when small actions yielded inordinately high results, but there were other times when high investments yielded little to no results. The result may be evaluated as a decent success using logical metrics and criteria.

6.3 Life as a Long Calculus

Life in both the academic and non-academic spheres are replete with daily decisions that, if plotted on a curve and extensively extrapolated as one would the path of a space shuttle, the resulting trajectory and final destination would be incalculable, granted the number of variables involved. All said, one has to begin the narrative somewhere.

I come from the land of educational mercenaries, because Ghanaian and other West African educators may be found anywhere from Australia to Zimbabwe (A to Z), working in content areas from archaeology to zoology, in mountain top to valley bottom higher institutions. Wherever there is research, teaching, and learning, one is likely to find a West African, from my experience.

Research products are necessarily the confluences and intersections of ideas. They are precipitates of matrices and combinations of ideological foundations. For the sake of deeper context, permit me to begin in Cape Coast, Ghana, where I was born. Cape Coast is known as one of the intellectual capitals of Africa for many reasons. Intellectual seats of the world are generally points of ideological intersections, the best example of which is the Mediterranean region of the world, connected to the Mediterranean Sea. It is instructive for the reader to understand that the term "mediterranean" is etymologically important, since it is derived from the two Latin words, "medi" and "terra" and translates as "middle of the earth." In his book, *The Great Sea: A Human History of the Mediterranean*, Abulafia (2011) illustrates the centrality of the Mediterranean Sea as being the nexus that connects major ancient civilizations in the sense that it not only provided travel (and hence ideological exchange and trading and war routes), but it also provided food and other means of sustenance. The mediterranean region was therefore the nexus of knowledge exchange across Africa,

Asia, and Europe, fostered by the surrounding water bodies for easy transportation of people and goods.

Per its very name, the geography of Cape Coast, my hometown, indicates that it is both coastal and a cape: a natural harbor. That geography laid the foundation for many European countries to have easy access to the local region for the exchange of ideas, and later, slave trade. In fact, Cape Coast and its neighborly town, Elmina, have two of the largest slave castles in the world. Those castles, interestingly, functioned not only as trading posts, but also, incidental and intentional educational centers: they were indeed sites for ideological exchange.

In the context of the United States as an example, locations of natural harbors became the sites for not only the easy transportation of African natives who had been captured and ferried across the oceans to become slaves, but they were also sites for ideological exchange. It is partly in the same vein that the skilled rice farmers of West Africa were actively captured and exported to the then-emerging rice farming plantations of the southeastern United States, such as Charleston, South Carolina and parts of Florida. This fact is well documented in Judith Carney's book, *Black Rice: The African Origins of Rice Cultivation in the Americas* (Harvard University Press, 2001) and Daniel Littlefield's book, *Rice and Slavery: Ethnicity and the Slave Trade in the Colonial South Carolina* (Louisiana State University Press, 1981). Similarly, African skilled metal workers were captured and exported to the Americas. Philip Reid is credited for his contribution in the casting of the bronze statue that sits atop the United States Capitol building. These references illustrate the positionality of Cape Coast as a major center of learning and intellectual exchange globally. It is therefore one of the first sites of education in the western tradition in Africa, and is the domicile of some of the most premier educational institutions for that matter. To be born in Cape Coast, Ghana, therefore, is a privilege in the academic sense, because this is town where even the ordinary people speak of themselves as having "pocket degrees."

6.4 Education in Ghana and Its Value in the World

A prophet has no honor in his own hometown, and so it is with Ghana's educational system—to stretch the analogy. I only discovered how good the quality of education in Ghana was—at least in my schooling years in Ghana—when I traveled outside the country and began interacting with other workers of the academy. That realization, however, was in confluence with my own identity evolution as an African academic in the United States.

By the time I began my academic career in the United States, I had been an Advanced Level (A-level) high school science teacher in Ghana for two years (1986–1988) and completed a learning fellowship in genetic engineering in Hungary (1988–1989). I then worked as a researcher in molecular biology at Georgetown University, my initial foray in the U.S. academic world.

Life in Hungary was very interesting in the sense that I was working in a large research institution (Biological Research Center), and my immediate co-workers could speak some English. I soon discovered that coming from an English-speaking country was a significant asset, because even in the then eastern Europe, English was important for publications in academic journals. I was therefore asked by a few colleagues to review their scholarly articles—mostly to check for English grammar. I was also employed by a professor of pharmacology in an international university to help him polish his English language skills. Since we were both scientists, our conversation revolved around scientific matters and mostly on pharmacology. I was relatively popular, because I was probably the best English writer available in my small circles.

By the time my foray in Hungary was over, I had not only lived with dozens of scientists from all over the world in an international hostel for scientists, but I had also discovered the value of a good quality education. That realization must have somewhat subconsciously fortified my sense of self-worth, although that was not enough yet to weather the professional terrain ahead.

6.5 My Initiation into the United States of America

My introduction to, and work at, Georgetown University was more of a surrealistic existence mired in culture shock, identity repositioning, and multiple magnitudes of personal evolution. By comparison, my scientific community in Hungary was rather laid back, only controlled by when our experiments called for our presence. Time was our friend, even though the environment was new. I was among a cadre of new fellows who arrived at the same time, and so I had company being new and rather green, lacking in even the basics of the Hungarian language. The wonder of waking up to a world totally blanketed by blazing white snow was shocking, because it was a total surprise, even if one had seen the pictures of snow in books. In good company, however, shared foibles were pardonable, because we all committed the same sins of butchering the Hungarian language and made fun of each other.

My initial encounter with the United States was a memorable one. An immigration agent would not believe that I was entering the United States with an H-1B (specialized) visa to work at Georgetown University, and so I was detained until my new boss could be reached to verify my identity after his class was over. In hindsight, I have to wonder what would have happened if he had not answered his office phone—because I am not sure he had a personal cell phone at the time in 1989.

The lack of knowledge as to how to prepare for a new academic (research) world, much less a new cultural environment, rendered me with some degree of rudderless helplessness. The speed at which I needed to adapt from a student-fellow in Hungary to a laboratory scientist in less than one week was dizzying at best, and discombobulating at worst. Coming directly from Hungary from a learning fellowship, I barely had the funds to purchase the basic provisions to begin my new life in the United States (from acquiring a shared apartment to where to purchase food and how things worked): even the electricity voltage was different: 240 V in Europe as compared to 120 V in the United States, thus rendering whatever electronic devices I had functionally useless. I had to learn fast to become a worker in an academic research environment to the best of my ability. No doubt, therefore, I still include that short period of time of initial acclimation to the United States as one of the most challenging phases in my life.

6.6 Identity Evolution: My K-12 Teaching Journey

My alma mater was not just a conventional university: it simultaneously prepared students to earn the conventional degrees in the content areas as well as prepare students to become licensed teachers, if they wished to pursue education as a career. With that as my educational background, I was lucky to arrive in the United States at a time (1989) when there was a shortage of STEM teachers (cf. Hutchison, 2012). It was therefore relatively easy for me to find a job as a science teacher, after I left Georgetown University. I therefore became a science educator after having worked as a science teacher in a K-12 institution, where I taught eighth grade physical science, basic ninth grade biology, and AP biology for ten years. The road to 10 years of my K-12 teaching experience was not an easy one. During my first year of teaching, I was almost fired for not knowing how to teach in the United States. My supervisor's evaluation was that although I had a good mastery of the content matter, I did not know how to manage my students in order to foster their learning. In retrospect, I agree with that assessment: I was expected to teach students living in the United States, not Ghanaian or African students, for whom I had been educated.

By the end of my ten-year tenure as a K-12 science teacher, I had learned several significant things about the connections between teaching and research, because I needed to research and relearn certain fundamentals in pedagogy. My future as a potential science education researcher was therefore organically charted. First, I learned that teaching is culture-dependent, and that to become a science education researcher, I needed to understand the cultural milieu of my future research context. Because I came from the traditionalist culture of the African classroom, I needed to more consciously re-learn the distinction among the different instructional methods and modes. In the traditionalist context, the instructional approach was largely lecture-based, in concert with laboratory work and limited discussion. The instructional environment was more authoritarian, reminiscent of the "sage on stage" and the maxim of students being "seen but not heard," unless called upon. If my research agenda was to be practical and tenable in the U.S. context, then I needed to embrace constructivism both as a philosophy and a pedagogical technique (cf. Marine, 2021).

6.7 My Academic Identity Evolution

One of the significant observations I made early in my professional career was that, whereas good writing skills were important in the academic environment, they were vital for optimal research productivity. In retrospect, I deem it tenable to argue that if the purpose of research is to add to the extant knowledge, then the ability to clearly delineate and articulate one's ideas with fine distinctions—in any language—confers a special advantage in extracting important and near-invisible knowledge from any data. This, in fact, is the reason for the concept of dissertation, whereby a research-based volume often akin to a book is written about just one subject or topic. Language and thought are two currencies that run in confluence (Vygotsky, 1986), and the stronger one's ability to use a language to explain finer details of a situation, the finer and clearer one's research products would be.

My master's and doctoral degree journeys were very smooth, partly because I came from a relatively competitive and even harsh learning environment in Africa. This stems from the fact that in emerging and traditionalist economies (comprising most countries in the world except for western countries), education is the easiest avenue for social and financial mobility. For this reason, education tends to be very competitive at each level. In the general educational market, the number of seats available in higher education—especially in the older, established universities—tend to be very few per capita, and the number of seats available for higher education becomes progressively smaller at each level of education. For this reason, the assessment systems tend to test for exclusion and selection (and therefore fulfil the selection function of testing (Lewin & Dunne, 2000).

To earn my zoology bachelor's degree during my schooling years, I had to conduct scientific research on the food web of a local lagoon, write a thesis, and successfully defend it. For several years in my university, there were only a handful of students who reached their final year as zoology students. As a successful zoology graduate of the University of Cape Coast, I only realized the difficulty of the Ghanaian education when I began my master's degree in the United States and discovered that my previous educational experience had been rather tyrannical, partly because it demanded a lot of intensive, competitive studies to become successful. By the time I began my doctoral degree journey, therefore, I had begun to become more confident in my academic abilities—largely because my Ghanaian education had provided me with a robust educational foundation across different subjects, and I was beginning to discover that although African countries were considered as "third world," African students had first-world intellectual capabilities.

African students often pose as an enigma to many people in the United States. The psychology of African students is different, because many of them (not all though, I must emphasize) almost necessarily come from families with relatively good financial backgrounds in their native countries. For this reason, they carry the psychology of confidence, which they transport across international borders. That psychology is often buttressed with strong educational preparation, because their families' economic means also translate into better educational preparation

(Blanden & Gregg, 2004). Most of these students have never discovered the perception of Africans outside their homelands. The result is that once in the United States or elsewhere outside Africa, they miss any signals of racial discrimination—for better or for worse: For better, because the knowledge of racialization and the related discrimination is a significant cognitive burden to carry on top on one's academic burden (cf., Hutchison's [2009] "minority effect"). It can therefore impair academic performance. For worse, because arguably, African students' oblivion to racism and racial discrimination cues often translates into non-realization of prejudicial mistreatments and racial discrimination. Ironically, this oblivion to prejudice and racial discrimination signals often results in various advantages, including not having to carry certain social burdens like being suspicious of what people from other races are possibly thinking when they are driving nice cars, or being educationally mis-assessed due to low expectations. For example, when I experienced challenges from my own students and even professors who were not aware of the variations in the English language across the world—including spelling and expressional differences—I had no reason to attribute their questioning to racism, because I was confident that I was speaking or spelling correctly. In time, however, I came to discover how variations of the English language and accents were perceived, assessed, and responded to, but that was a realization to come.

6.8 The Journey to a Doctoral Degree

By the time I began my doctoral studies, I had lived in the United States for about eight years and had earned a master's degree. During the master's program, I encountered professors who were not only nice and even hospitable, but fair to me. That notwithstanding, I also could not help but notice that some were rather surprised by my good academic performance that led to a *summa cum laude* designation in my degree. Granted this intermixture of curiosity and courtesy, I became convinced that people in general are nice in personal disposition, but that some did not have the necessary information to overcome the broader society's racial bias, and that when provided with more information, their racial curiosities could convert into logical conclusions.

Being an immigrant is a condition that compels specific questions and apprehensions, which in turn bear on one's evolving identity. Important as they are in shaping one's emerging life in the new country, the hurriedness and the tedium of one's new life—often mired in survivalist tendencies—leave little mental room to seek answers to questions that often are necessary but unconceived—because one has no frame of reference to begin asking those questions. This is akin to a new, rudderless teenage mother who is trying to raise the new baby. Mistakes become inevitable, unless good social support is available and accessible.

There is a philosophical largesse which posits that the phenomenon of time is a gift that stretches life's events, so that its consumers are not overwhelmed by the compression of events. Over time, I learned to disregard a lot of advice I was offered by people, for a couple of reasons. First, I had had the opportunity to make my own observations and reach the conclusion that Africans are often misunderstood: that, because we hailed from "developing countries," our intellectual abilities were also "developing." For this reason, I had learned not to calibrate my abilities with external signals. Another observation I made was that, even if I spoke standard English with a decent level of mastery, my African accent was frequently misperceived and misconstrued with intellectual suspicion. For these reasons—and having completed a master's degree in the United States—I suspected that I could speed up the Ph.D.-acquisition process by creating and living in my own micro-universe. I was able to take multiple courses while working full-time and yet completed the program, including my dissertation, so quickly that I was asked to wait an extra semester before the final defense, and still won a dissertation award.

I believe in a higher power and her providence in nature. I also believe in logic and phenomenology. For this reason, I subscribe to the philosophy that summates as, "Pray as if it all depends on God, but work as if all depends on yourself." I believe that this philosophy helps to generate "good luck" through self-fulfilling prophecy. Phenomenologically speaking, a smiling person who exudes gentility of spirit and bright countenance is more likely to induce a reaction in kind—and that is one avenue for inducing good luck in social interactions.

By sheer luck, blessing, or otherwise, I was given a beneficent male in a doctoral advisor, who was instrumental in helping me to discover that I was student of good promise. He subscribed to the humanism school of thought, and so he had a refined sense of empathy. He was also a conscious, master teacher and a globalist, before secular globalization concepts congealed in the 1980s. Being a foreigner, he must have taken an interest in me, as he had other foreign doctoral students that he had supported as a doctoral advisor and mentor. For those reasons, I only later discovered that I was inadvertently being cultivated into a globalist by way of research, and a humanist by way of educational philosophy.

6.9 Consciousization of Philosophical Landscapes

So far, it has been noted that education in developing countries is qualitatively and quantitatively different from what is practiced in an advanced economy like the United States. Whereas developing economies are more traditional in their educational traditions and respect hierarchical structures, the western educational environment is more meritocratic. This also means that professors are viewed through a lens of superiority, and that correlates with their practice of the "sage on stage" mode of instruction mentioned earlier. It was customary of them to be rather tyrannical in their philosophical outlook on instruction, and concomitantly sometimes had unreasonably high expectations of students—which expectations most of them could probably not meet themselves, from my personal judgement. This was further reflected in the assessment philosophy whereby, in my college years in Ghana, 40–50% was deemed a passing grade, depending on the course and the instructor. Many professors enjoyed being known as the ones whose courses were difficult to pass. On the other hand, I found western educational philosophy to be more humanistic. Course syllabi clearly delineated not only the course content, but the course expectations and even the grading scheme—something that was mostly absent in my African education.

Because I was used to the oft-tyrannical educational demands in Africa, I found the United States educational environment to be a very pleasant one in which to function not only as a student, but also, as a researcher. Here again, it is vital to note that the value of a good doctoral advisor cannot be overemphasized. My advisor used to organize summer workshops for the local science teachers and summer instruction for master's students, and all his doctoral students served as the interns. In hindsight, this was the genesis of my unconscious aspiration to become a professor. During the summer internships, we co-taught the courses and observed our advisor teach the students. We heard him advise students and were even allowed to sit in advising sessions with students (when the advisement was academic and did not pertain to personal matters). We graded papers and were required to learn to work online—and this was between 1998 and 2000—at a time when very few instructors were beginning to learn to use the internet for instruction. In fact, my doctoral advisor also had an Eisenhower grant for conducting workshops for science teachers on the use of the internet in science instruction during that time period. As a part of our internship, my doctoral advisor had identified a few master teachers in an urban middle school, and we were made to conduct observations of these master teachers in action and discuss their processes.

I deem these internship opportunities as very vital to my personal ascension into the academy, because on one occasion I was asked to research the importance of diversity and inclusion considerations in science teaching. When I completed the instruction, I was showered with praise from my advisor (some degree of which, I believe, was more an encouragement, and not my actual performance). That notwithstanding, my confidence was boosted by the encouragement, and I began thinking about including that topic as a special research mastery of mine. A few years later, when I began my formal work outside my doctoral institution as a tenure track professor, that became one of my anchor courses for which I eventually wrote several books and articles (more on this below). In consonance with this point, my doctoral advisor—who himself was an author of several books—informed me that I had good writing skills. Soon thereafter, my advisor asked me if I would be interested in a position as a Clinical Instructor of Science Education. This was an exciting opportunity, and so I responded in the affirmative.

6.10 Discovering the Nature of the Academic Landscape

During my youthful years in Ghana, my father worked at my alma mater, the University of Cape Coast in Ghana. I therefore virtually grew up on the university campus and heard many stories about life in the academy. That notwithstanding, I quickly learned that the United States university culture and traditions were very different, with one exception: the humans who operated within them behaved similarly: there were always wars and rumors of war among people, and there were behaviors that made one wonder if certain people disliked you or were advocates for you. I also learned that there was such a thing as "guilty by association," à la the French expression, "*Les amis de mes amis sont mes amis*" (the friends of my friends are my friends). By extrapolation therefore, I learned that, potentially, the adversaries of one's friends may not be people on whom to depend for one's professional mobility—especially if they are in positions of power.

6.11 Near-Mistakes and Lessons Learned

It is no sin to make a mistake, but it is a sin if you do not learn from your mistakes. The scarcity of space will not permit me to discuss many of the lessons I learned during my early life in the academy. In the next sections, however, I will focus my discussion on the more salient lessons I learned on my way to ascending the academic ladder as a teacher, researcher, and a journal editor.

Advice #1: Prepare yourself, trust in your abilities, and leave your alma mater

While learning to become a tenure-track professor, I was fortunate to encounter a few individuals whose advice was pivotal in my eventual success in the academia. The primary one was my doctoral advisor, whom I mentioned earlier in this essay. Having been a K-12 teacher in the Atlanta, Georgia area for ten years and being comfortable with the local terrain, I had become obsessed with finding a tenure track position in the Atlanta metro region. To this end, I was almost certain that I would be considered for one of the three positions that opened up in my department—after all, I was then temporarily running a large program in science education when my advisor retired, and since I was his only assistant at his retirement, I believed that I was a shoe-in for the position. Imagine the surprise, therefore, when I was ushered into a meeting and was informed that I would not be considered for the position, because I was a Ph.D. product of the same program. (If I had been hired by my graduating university, that would have been considered as "academic incest".) I must note, however, that someone in my same situation was hired in the same department, but not in the same unit. That made me question the decision, because I was all but certain that I was the automatic hire. In my bewilderment, I sought the advice of my trusted doctoral advisor, who revealed a secret to me: If possible, never work in the same institution where you earned your doctoral degree. He informed me that if I stayed, I would always be the subordinate of the same people who taught me in the program. Besides, I later realized that home-grown candidates are often limited to clinical positions that are exclusive of permanent tenure. He encouraged me to apply to outside institutions where my true abilities would speak for me and avoid the possible identity of a favored student of the program. He also assured me that given my performance under his tutelage, I would be an attractive candidate elsewhere.

The lesson is that unless you have compelling reasons to stay, avoid seeking a position in your graduating university of your doctoral program (especially in the same department), because you are unlikely to best your former instructors and mentors who "made" you who you are. They probably know more than you, not only in content matter, but also the history of the department. In my own experience, I was naively comfortable at my doctoral institution, only to realize that I had no idea how departmental politics worked. Although I believed that I was the logical shoe-in for multiple faculty positions in the department, I reluctantly and rather timidly followed my mentor's counsel and decided to seek positions in other institutions.

Advice #2: Trust yourself and trust the process

Having been given the proverbial shot in the arm by my advisor, and yet not knowing my own capability, I set out to apply for a few positions. Surprisingly enough, I began getting job interviews! Because my family wanted to live in a racially-diverse town, I decided to bring along my spouse so as to drive around the town to gauge how friendly it would potentially be for our children—especially around the local schools.

Job search processes are often replete with self-doubt and self-limitation, and candidates of color are more likely to engage in self-exclusion because of self-doubt, especially if they have not had anyone inform them of the imposter syndrome (e.g., Collins et al., 2020) whereby newly-minted doctoral graduates do not feel adequately prepared to engage the real world of the academia (cf. Dancy & Jean-Marie, 2014). This is why the treasured encouragement of my advisor was important in my decision as to what kinds of position to apply for.

During the job search, I quickly secured campus invitations to two, and then three, then four institutions. It became clear to me that having a doctoral degree in science education was an advantage, because it was assumed that we could concurrently double as STEM educators, especially in the relatively smaller institutions. Because the search committees had different campus interview schedules, I quickly learned not to immediately commit to the first offer I received. This was important, because I received my first offer within 24 h of my campus interview. The salary was fair for the local cost of living; however, the town was not as diverse as my family needed it to be. I informed them to give me a few days to make my decision, partly because within the next couple of days, I had another campus interview in another institution that was closer to Atlanta, my then preferred city.

My second interview resulted in a better offer, and so I called the first institution to decline their offer. Once again, I had another campus interview scheduled with more invitations coming, and so my confidence level was slowly increasing, although filled with high anxiety. Again, I asked for a few days to make my decision while I visited my third campus for interview (and yes, still with my family in tow for all three interviews so far). My third interview was at The University of North Carolina at Charlotte. While at the interview, the search committee arranged for a real estate agent to show my wife around Charlotte and some housing options. The search committee hosted my family, including our then two little children for a nice dinner. By the end of the dinner, I had found a new home: I would accept the offer if given, and they did!

Somewhere along the line, I had learned the popular wisdom that the job interview was a two-way street: you are interviewing your interviewers as much as they are interviewing you. During the last job interview, all the signals were present that this institution would not only be a great place in which to work, but that my department colleagues would be great supporters of my research agenda and professional trajectory. This paragraph circumscribes my professional life for over 22 years at the time of this writing. The lesson here? Trust in your abilities if you have adequately prepared yourself for the job market and expect good tidings accordingly.

Advice #3: Be a big fish in a small pond

Early in my career as a science educator, I encountered a fellow alumnus of my doctoral institution. While commiserating about our alma mater's news and memories of mutual acquaintances, he revealed a nugget of wisdom that made a lot of sense to me: "Be the big fish in a small pond." The pond may be definable as any environment, be it one's place of work, a professional organization, or even a journal review board. Somewhere along my professional journey as an educator, I learned that many high school athletes were strategic in their high school selection: A small school may not have the financial capacity to support a serious athletic program, but a very large school may have too many good players for you to have the opportunity to play and shine; a medium size high school, however, will likely have the resources to support various athletic programs, but small enough to allow you to play. The lesson is that it is much easier to ascend to the executive committee of a smaller organization than a larger one where the pillars of the industry are vying for visible positions.

The preceding advice is even more important when one considers the fact that your early rise through professional organizations before tenure is much more important in determining your achievement of tenure (by way of service requirements) than post-tenure, when your survival in the profession is less dependent on such perceptively higher achievement. Notably, the same professional power players of the national level often attend the professional meetings at the smaller, regional level, and will be compelled to know who you are, especially if you hold a visible position at that level. In conclusion, find a pond that is large enough to support your interests, but small enough to allow you to play.

Advice #4: Being a professor is a lifestyle, not a job

During my doctoral program, one of my professors stated that becoming a professor was a lifestyle, not a job. She was a science educator who was also a practicing, part-time pharmacist. She loved being a professor, because she was able to control her time and lifestyle, and with a modicum of discipline, she could control her own destiny and happiness. That notion was intriguing to me, because I interpreted it to mean that there were no limits to what I could achieve if I invested time into my professional capabilities and disciplined myself. As I had noted earlier, I was never an ambitious person; however, I discovered that I can be very disciplined when needed, and did not like to be professionally outcompeted or left behind, unless I voluntarily allowed it and justified it. The amalgamation of this personality or mindset, however, is only permissible when one consistently engages in several years of reflection. I was compelled to do so, because my African immigrant positionality necessitated the same, and defined the axes thereof: Within the same frame of mind, I had to learn to concurrently accommodate the conversion of the Ghanaian currency into dollars as well as convert my African pleasantries through U.S. cultural filters and translate them into pedagogical apparatus. This point is parabolically-stated to induce the flexibility of interpretation and to offer the economy of verbiage.

Advice #5: Uncover what comes easy to you and pursue it

In physics, efficiency is definable as the output per unit time versus energy input. To achieve true efficiency in life, however, it is vital to account for emotional investment per output. In academia—especially in larger, research-oriented institutions—the notion of "counting beans" often dominates the zeitgeist. The safest avenue to achieve tenure is to ensure publications in reputable journals and publishers, and in as many as possible. In this vein, I realized very early in my profession that in addition to writing journal articles, I could employ my writing skills to write as many books as possible; after all, with a little discipline, one written page a day would easily translate into a 200-plus page book per year. In retrospect, book authoring was an idea that was inculcated, once again, by my doctoral advisor and mentor, who himself was the author of multiple textbooks of science education. After my defense of my doctoral degree, he advised me to consider publishing my dissertation. I followed his advice, and the rest is history.

Another important wisdom I learned early on in my career was the idea that "if you practice what you love, you will never work a day in your life." Being an immigrant who had lived in Africa, Europe, and the United States, and then becoming a science educator, I observed that there were natural educational impediments that one encounters when migrating across national borders. As it turned out, there was little formal research that had hitherto been conducted on the subject. That became the subject of my dissertation, my first book (Hutchison, 2005), and several journal articles (e.g., Hutchison et al., 2005). Because I was an immigrant who was therefore an "international" personality, I expanded my area of research into the adjacent subject of globalization once I had exhausted my initial interest in the subject of academic migrant workers (e.g., Hutchison, 2016, 2022; Wiggan & Hutchison, 2009). Being an immigrant also naturally qualified me as someone who could work in the broader subject area of diversity and even authored a textbook in the discipline (see Hutchison, 2011). Once I had gained more confidence in the academy and had developed more profound insights in the field, I viewed myself as a philosopher-at-large. I began tackling broader issues in the field of education and developed some theories, such as the concept of *cultural constructivism* (Hutchison, 2006), the *minority effect* (Hutchison, 2009), and *disability-by-induction* (Hutchison et al., 2014). All these topics related to issues that directly impacted my personal life; for this reason, the issues were not only interesting, but also, I was automatically a pseudo-authority on the topic. It was clearly an advantageous professional strategy for reaching the apex of the academic ladder.

Advice #6: Follow the tricks of the trade for accelerated productivity

For those on tenure track whose responsibilities include knowledge production, but whose work output is assessed by way the number of articles, book chapters, and grants, to mention a few forms of academic outputs, I discovered that there are ways to become "perceptively" successful and highly efficient—if you can create a productive network. Although I personally did not engage in this strategy, experienced professors are familiar with what one may call the multiplication-by-group-effort concept. In this strategy, each person in a group of three professors (as an example) can cultivate a topic, and the other two colleagues are made the co-researchers and co-authors in each other's project. The result is that each professor now has a first authorship, a second authorship, and a third authorship for a total of nine authorships among the group of three. It is, indeed, one of the open secrets in the academic world: the strategy of collaboration and induced efficiency.

The problem for professors from underrepresented groups in science education used to be that our research topics used not to be interesting to our colleagues; that is why collaboration with other colleagues within and outside your institution is important. Another problem is that your colleagues may not necessarily be socially interested in you, as a person, because, like it or not, academic success is often a function of one's social network at the local, national, and even international levels. In that sense, the intentional cultivation of social networks while in the doctoral program is very important for future success in the academy. Beyond that, take advantage of mentoring and other networking opportunities offered at various conferences and in the university.

6.12 Conclusion

All roads lead somewhere, but a predetermined journey will lead its undertaker to a specific destination. Success in any profession is a series of events that are fueled by intention, focus, and determination. My personal journey as a science educator and researcher has taught me that it is almost impossible to fail in the profession if the processes outlined in this chapter are followed, along with planning one's social life in ways that harmonize with one's life as a professor. Science educators understand basic engineering concepts. I view life as a canvas that presents us with the opportunity to engineer our lives. Great machines are so considered, not only because they can generate good products, but also because they are efficient. An efficient, successful life has less friction and therefore generates less heat and uses less energy: it is emotionally efficient. Our social and professional lives are intertwined, and good life engineering principles naturally harmonize them into the best possible life.

References

Abulafia, D. (2011). *The great sea: A human history of the mediterranean*. Oxford University Press.

Andelin, A. (1994). *Man of steel and velvet*. Pacific Press.

Blanden, J., & Gregg, P. (2004). Family income and educational attainment: A review of approaches and evidence for Britain. *Oxford Review of Economic Policy, 20*(2), 245–263. https://doi.org/10.1093/oxrep/grh014

Carney, J. (2001). *Black rice: The African origins of rice cultivation in the Americas.* Harvard University Press.

Collins, K. H., Price, E. F., Hanson, L., & Neaves, D. (2020). Consequences of stereotype threat and imposter syndrome: The personal journey from STEM-practitioner to STEM-educator for four women of color. *Taboo: The Journal of Culture and Education, 19*(4), 161–180. Retrieved from https://digitalscholarship.unlv.edu/taboo/vol19/iss4/10

Dancy, T. E., & Jean-Marie, G. (2014). Faculty of color in higher education: Exploring the intersections of identity, impostorship, and internalized racism. *Mentoring & Tutoring: Partnership in Learning, 22*(4), 354–372. https://doi.org/10.1080/13611267.2014.945736

Hutchison, C. B. (2005). *Teaching in America: A cross-cultural guide for international teachers and their employers.* Springer.

Hutchison, C. B. (2006). Cultural constructivism: The confluence of multiculturalism, knowledge creation, and teaching. *Intercultural Teaching, 17*(3), 301–310.

Hutchison, C. B. (2009). The minority effect as a human phenomenon and its educational implications. In C. B. Hutchison (Ed.). *What happens when students are in the minority: Experiences that impact human performance* (pp. 3–17). Rowman and Littlefield.

Hutchison, C. B. (2011). *Understanding diverse learners: Theory and practice.* Copley Custom Textbooks.

Hutchison, C. B. (Ed.). (2016). *Experiences of immigrant professors: Cross-cultural differences and challenges, and lessons for success.* Routledge.

Hutchison, C. B. (2022). Globalized science education through technology in a multicultural education context. In M. Atwater, (Ed.), *International handbook on multicultural science education* (pp. 799–817). Springer. https://doi.org/10.1007/978-3-030-37743-4_30-1

Hutchison, C. B., Butler, M. B., & Fuller, S. (2005). Pedagogical communication issues arising for four expatriate science teachers in American schools. *Electronic Journal of Science Education, 9*(3). http://wolfweb.unr.edu/homepage/crowther/ejse/ejsev9n3.html

Hutchison, C. B., Campbell-Whatley, G., & Wiggan, G. (2014). Disability by induction: Academic under-performance when out of context. *Insights on Learning Disabilities, 11*(1), 1–8.

Hutchison, L. F. (2012). Addressing the STEM teacher shortage in American schools: Ways to recruit and retain effective STEM teachers. *Action in Teacher Education, 34*(5–6), 541–550. https://doi.org/10.1080/01626620.2012.729483

Lewin, K., & Dunne, M. (2000). Policy and practice in assessment in Anglophone Africa: Does globalisation explain convergence? *Assessment in Education, 7*(3), 379–399.

Littlefield, D. (1981). *Rice and slavery: Ethnicity and the slave trade in the Colonial South Carolina.* Louisiana State University Press.

Marine, M. M. (2021). Translating constructivism into pedagogy from philosophy to practice: Active project-based learning. *The International Journal of Humanities Education, 19*(1), 39–51. https://doi.org/10.18848/2327-0063/CGP/v19i01/39-51

Vygotsky, L. (1934/1986). *Thought and language.* MIT Press.

Wiggan, G., & Hutchison, C. B. (Eds.). (2009). *Global issues in education: Pedagogy, policy, practice, and the minority experience.* Rowman and Littlefield.

Chapter 7
The Community is My Laboratory: Leveraging My Faith Walk to Reconceptualize Service as Scholarship in the Academy

Natalie S. King

To love science and to know God.

To be in the world, yet not of the world.

To work in schools while not "doing school."

To be in community with the marginalized while placing them at the center.

Or I can take a more anthropological perspective, making the strange familiar and the familiar strange.

These are the dichotomies that make me complex yet so uniquely beautiful.

This is my journey into science education and academia. Unorthodox yet necessary. This is my story; this is my song.

Life, well, it hasn't been easy. It's had its share of ups and downs.

The struggles, the successes, the trials, the triumphs, the valleys, and the mountain tops, they make the road trip scenic, and the journey worthwhile.

In this chapter, I choose not to solely focus on my work because work is only a byproduct of the person. I want to share the meaning of being, becoming, and evolving as an early career researcher in this space.

N. S. King (✉)
Georgia State University, Atlanta, GA, USA
e-mail: natalieking@gsu.edu

M. M. Atwater and B. A. Garvin (eds.), *Creating and Maintaining a Home in Science Education Research*, Contemporary Trends and Issues in Science Education 65,
https://doi.org/10.1007/978-3-031-99382-4_7

7.1 My Faith Walk with Ordered Steps

My faith walk with God is the foundation of my life. In writing this chapter, I felt conflicted about how to share my journey as a science education researcher because while it's been rewarding, the turbulence has tested my resolve, challenged my identity, and deepened my purpose. The path that I have traveled has been filled with ups and downs, ebbs and flows, yet one truth remains clear: it was God who led me here. Still, my journey did not follow the traditional path of a tenure-track professor. Instead, God called me to work directly in the community, leveraging evidence-based practices to make a tangible impact. For me, it was never about becoming the most highly cited scholar or being internationally recognized for theoretical and methodological contributions. My driving force has always been action and creating real change by serving those who have been disenfranchised. I am a humanitarian at my core, called to uplift, inspire, and use science as a vehicle for liberation and justice.

This calling is not of human origin. It is God who has placed me here, to be a beacon of light in dark places, shining for the world to see. While reflecting on Acts 5, I was particularly moved by Gamaliel's wise counsel to leave the apostles alone: "*If their purpose or activity is of human origin, it will fail. But if it is from God, you will not be able to stop these men; you will only find yourselves fighting against God*" (Acts 5:38–39). That passage resonates deeply with me because it reflects my journey. My approach often defies the conventions of the tenure-track world, yet God has honored my obedience. It is through Him that I have experienced success, and it is all for His glory.

As I reflect on my journey, I am reminded of the words of Jesus: "*By their fruit you will recognize them*" (Matthew 7:16). Our lives as disciples of Christ must bear fruit that reflects God's love, justice, and purpose. It is not enough to claim faith; our actions must demonstrate the transformative power of that faith. This fruit is evident in the work I do in science education by uplifting communities, creating access and equity in spaces where Black and Brown children have been marginalized, and ensuring that my labor reflects the values of Christ. Just as Gamaliel recognized that God-given assignments cannot fail, I trust that the fruit of my obedience to His calling will testify to His power and glory.

With this grounding, I share lessons that I learned along the way. My work as a science education researcher has found its home in service to the community. My laboratory is not confined to academia but is alive in spaces where equity, access, and justice desperately need to reside. In this work, I am reminded daily to step outside of tradition to fulfill God's purpose for my life in the earthly realm. God knows who I am, He delights in my ways, and orders my steps.

7.2 Finding a Home

At the heart of my research and praxis is my love for people and the many places we live and call "home." Community-engaged scholarship bridges the gap between academia and community, fostering reciprocal relationships that not only address systemic inequities but also empower communities through collaboration and co-creation of knowledge. In the context of science, technology, engineering, and mathematics (STEM) education, such engagement is critical for providing access and equity to populations who have been marginalized. Community-engaged work in STEM education emphasizes the significance of context and culture, demystifying STEM fields for historically excluded groups and dismantling barriers to participation (Brown et al., 1989; McGee & Bentley, 2017). By centering community voices and addressing systemic inequities, this approach creates transformative opportunities for engagement and learning that resonate deeply with the lived experiences of communities that have been historically relegated to the margins.

Place, in this context, transcends the physicality of geographical location to encompass what bell hooks (1990) describes as a *homeplace*: a site of resistance and refuge where marginalized individuals find solace, healing, and the strength to challenge oppressive systems. hooks (1990) argues that *homeplace* serves as a foundation for resistance, particularly for Black women, where the personal and political intersect to foster radical transformation. These spaces, imbued with cultural memory and resilience, become critical in resisting sociopolitical and capitalist agendas that seek to undermine Black communities.

This intentional and strategic connection between space and political resistance has informed my scholarship and positionality as a pre-tenure and now tenured Black woman faculty member. In *Teaching to Transgress*, hooks (1994) noted that education itself can be an act of liberation and a means of reclaiming agency and disrupting systems of domination. Similarly, McGee (2020) highlights the dual burden and potential of Black women in academia to resist systemic oppression while fostering equity and justice within and beyond institutional spaces.

In this place, I have cultivated an antipatriarchal, antiracist, radical feminist commitment to challenging the sociopolitical structures that perpetuate inequities. My work interrogates traditional notions of the professoriate and STEM education, reframing them within a decolonial praxis that centers the humanity, creativity, and agency of Black communities (Tuck & Yang, 2012). This decolonization process is transformative for me personally and the individuals I have the honor of serving as it reinforces my commitment to liberation and racial justice.

My situatedness at the nexus of community, scholarship, and resistance helps me to embody a vision of academia that is deeply rooted in love, justice, and the collective liberation of Black and Brown communities. As McGee and Bentley (2017) assert, STEM education and academic spaces must shift toward frameworks that prioritize equity, identity, and the dismantling of systemic barriers – a charge I hold as both a responsibility and privilege in my role as a scholar.

7.3 Welcome Home

Deciding on a starting place to tell my story is always a challenge because there are so many lessons and experiences that I could share. Growing up in church, participating in Bible quizzing, and internalizing the idea that "this world is not my home; I'm just a-passing through" formed the moral and ethical foundation that would guide my personal and professional journey. One significant event that marked a pivotal shift in my life occurred during my master's degree program when I enrolled in an Inquiry-Based Science course taught by Dr. Rose Pringle—a Jamaican Black woman science educator and tenured professor. At the time, I was a full-time high school chemistry and biology teacher and was simultaneously preparing to retake the Medical College Admission Test (MCAT) with the goal of applying to medical school. My love for children and science shaped my aspiration to become a pediatrician...or so I thought.

As an undergraduate participant in the Ronald McNair Scholars Program, I was introduced to applied research, which piqued my interest in health disparities. I had considered the joint MD/PhD program to address these inequities in Black communities. Through local outreach initiatives, I engaged in efforts to reduce preventable diseases such as hypertension, diabetes, and cardiovascular disease because these ailments disproportionately plagued Black communities. I worked alongside medical professionals to deliver health education, promote early screenings, and support lifestyle interventions tailored to the cultural contexts of the individuals we served. These experiences underscored the systemic barriers to healthcare access and the urgent need for culturally responsive interventions that empower communities to take charge of their health. Additionally, I gained insight into the clinical, academic, and outreach aspects of medicine, shaping the foundation for my current work in STEM Education.

Dr. Pringle's course challenged me to think about science teaching and learning in innovative ways. After completing my portfolio project, she invited me to meet on campus to discuss my work. During that meeting, she asked a life-changing question: "Have you considered a PhD in Science Education?" I was perplexed and had so many questions. I wanted clarity on the process to earn a PhD and career possibilities after completing the degree program. While I had previously applied to MD/PhD programs and considered a PhD in epidemiology as an entry point to conduct research in community-based settings and work toward mitigating health disparities in Black communities, I had never considered pursuing educational research—let alone focusing on K-12 science. Dr. Pringle explained the process in detail, highlighting how my passion for engaging Black girls in science could evolve into a research agenda. She recognized my natural ability to create culturally relevant curricula that resonated with my students' lived experiences and interests. Her affirmation of my potential and careful guidance catalyzed a transformation in how I envisioned my career and purpose.

In 2012, I began the PhD program in Curriculum and Instruction with a specialization in Science Education at the University of Florida. Immersing myself in literature, I encountered transformative concepts such as culturally relevant pedagogy (Ladson-Billings, 1995), culturally healthy children (Ladson-Billings, 1989), and the notion of White privilege (McIntosh, 1988). These frameworks challenged my thinking about education and systemic inequities, prompting deep reflection on my role as an educator. As I engaged in seminar courses with other doctoral students, we grappled with critical theories that illuminated the entrenched structures of oppression in education. I delved into critical race theory (CRT) (Bell, 1992; Delgado & Stefancic, 2017), which provided a lens to interrogate racism as a permanent fixture in U.S. society, and explored the concept of hegemonic structures (Gramsci, 1971), which revealed how power operates to maintain inequities within education systems. The deeper I immersed myself in these theories, the more I realized that educational systems were not broken as I had once assumed; rather, they were functioning precisely as designed to perpetuate disparities that kept certain groups disenfranchised (Gillborn, 2005).

Alongside these theoretical foundations, I discovered qualitative research (Creswell & Creswell, 2017; Flick, 2007), which opened my eyes to the power of narrative and context in addressing questions of access and equity in science education. Through qualitative methodologies, I began to explore the lived experiences of Black girls and science teachers in ways that centered their voices and humanity. Narrative inquiry and storytelling became essential tools in my pursuit of justice-centered education (Clandinin & Connelly, 2000). This exposure not only shaped my research agenda but also deepened my commitment to creating joyful and equitable learning spaces where Black students could thrive.

Even while pursuing my doctoral degree, I remained deeply engaged in my local church. I led the children's choir, participated in praise dancing, and assisted with educational programs such as the Black History Brain Bowl and summer camp. These experiences reinforced my belief that faith, community, and education were intertwined and should not be compartmentalized. My church involvement provided a grounding force, reminding me that my work was not merely an academic exercise but a calling rooted in service and justice (Emdin, 2016). It was through this balance of academic scholarship and community engagement that I found clarity in my purpose, to create transformative learning experiences that uplifted youth and their families.

Currently, my research is deeply rooted in qualitative approaches through narrative inquiry and storytelling, with a focus on advancing Black girls in STEM education, community-based STEM programs, and the role of curriculum in fostering equity in science teaching and learning. My scholarship aims to dismantle barriers and reimagine STEM education as a space of joy, belonging, and empowerment. I aspire to leave behind a legacy of humanizing practices that affirm the identities and brilliance of students who have been historically disenfranchised and marginalized.

It was a Black woman science educator who first recognized the science educator in me. She saw the talents and exploits that I could accomplish in this field, driven by my passion for creating joyful learning spaces for children while simultaneously addressing inequities in a movement toward justice. I strive to do the same by recognizing, affirming, and cultivating the brilliance of Black students who deserve to see themselves as scientists, innovators, and change-makers.

7.4 Turning a House into a Home through Research

Upon graduating with my doctoral degree, I took a tenure track position at Georgia State University and moved to the metro Atlanta area with a clear research agenda focused on (1) advancing Black girls in STEM education, (2) establishing and sustaining community-based programs through research-practice partnerships, and (3) fostering equity in science teaching and learning through curriculum development and teacher diversity.

Almost 10 years into the profession, I now serve as a tenured Associate Professor of Science Education, disseminating my scholarship through research presentations, invited keynote talks, workshops, and publications in peer-reviewed and high impact journals. I have also written a book, *Let the church say amen to STEM: Guidebook to launching and growing extraordinary youth programs*, that speaks to faith-based leaders on how to engage their congregants and local community in meaningful ways to promote STEM Education (King, 2019). My publications demonstrate a unique ability to contribute intellectually to the knowledge base in my field, with influence that extends to practitioners and the larger public. Furthermore, as an emerging scholar in science education, my research has been recognized by the National Science Foundation (NSF) with nearly $5.5 million in external funding across my research areas. I discuss my approach to crafting an exciting research agenda through my scholarly interests and grant projects.

Advancing Black Girls in STEM Education: My research challenges the capitalist and workforce development agendas that encourage Black girls' engagement in STEM education, reframing the STEM disciplines as a mechanism to promote personal development, academic growth, and social justice. I leveraged narrative inquiry and critical race methodology to produce counterstories of Black girls' informal and formal STEM learning experiences (King & Pringle, 2019). I have also explored the characteristics of teachers who promoted Black girls' academic success and engagement in STEM learning (King, 2017). Furthermore, I serve as an associate editor for the *Journal of African American Women and Girls in Education,* where I support the review and advancement of research concerning the educational experiences of Black girls and women. In 2020, I received the most prestigious honor from the National Science Foundation for junior faculty—the Early CAREER Award (# 1943285) with $1.1 million in initial funding and an additional $240,000 supplement to conduct a five-year longitudinal mixed methods study investigating Black girl brilliance and STEM identity development. This interdisciplinary study explores how Black girls exude brilliance in STEM learning spaces and their intent to enroll in college and pursue a STEM trajectory. I am spearheading research to re-author the positioning of Black girls in formal and out-of-school spaces and currently partner with an all-girls public charter school and nonprofit organization that provides wraparound services for girls.

Establishing and Sustaining Community-Based Programs through Research-Practice Partnerships (RPPs): As a former high school science teacher, I recognize the inequities and disparities in course offerings, qualification of teachers, and access to resources. If our nation is truly committed to broadening participation in STEM, we must consider out-of-school learning spaces. My commitment to communities and active engagement in RPPs transcend all areas of my scholarship. I have directed out-of-school STEM programs as founder and executive director of *I AM STEM* for nearly 15 years, which has served nearly 5,000 K-12 children across 23 U.S. states, Canada, the Caribbean, Ghana, and South Africa. In 2018, *I AM STEM* received the Rose Duhon-Sells Program Award from the National Association for Multicultural Education (NAME) in recognition of our commitment to multiculturalism and preparing students to become active citizens in a democratic society.

I AM STEM has a track record of providing transformative and anti-racist STEM curriculum for Pre-kindergarten—12th-grade students that decenter mainstream pedagogical practices and embrace students' histories and identities as the vehicle to drive instruction. We introduce children to concepts such as cosmetic chemistry, artificial intelligence, cybersecurity, esports, and gaming and design, intentionally connecting the curriculum to STEM careers. We acknowledge the cultural wealth that exists in communities that we serve to actualize equitable STEM learning experiences. I also led a research team exploring determinants of Black families' access to this community-based STEM program using a latent class analysis, which is now published in *Science Education* journal (King et al., 2021). The latent class model extracted three distinct classes. Our findings revealed that all parents in the program were likely to enroll their elementary-aged children, but the likelihood of enrolling middle or high school children increased with parents who were college-educated. This study also revealed that, regardless of income or neighborhood lines, Black parents of all social classes and family dynamics learned about the program through information sharing. Furthermore, the program's location increased access to informal STEM learning opportunities for parents who experienced poverty, while the program's curriculum attracted middle-class Black families who were not residents of that particular community. Another distinctive feature of *I AM STEM* is our collaboration with faith-based institutions to position them as catalysts for expanding access to STEM programs. Churches play a vital role in fostering academic excellence and leveraging skills to create social and economic opportunities for their communities. To support holistic development, we must cultivate a 21st-century Christian culture that uplifts both congregations and the broader community.

Role of Curriculum and Teacher Diversity in Fostering Equity in Science Teaching and Learning: My third line of research is not mutually exclusive from the others but foregrounds my role as a science teacher educator. I served as the Principal Investigator on a $3 million NSF Noyce Track 2 Award (# 1852889) seeking to diversify the STEM teaching workforce in high-need school districts with a specific emphasis on Black males. The Noyce project had major goals of recruiting and preparing STEM professionals to teach in high-need secondary schools and engage them in robust and innovative professional learning experiences for their development as teacher leaders. My colleagues and I partner with two large school districts and nonprofit organizations, such as the 100 Black Men of America, to attract, prepare, and retain highly qualified and diverse STEM teachers in metro Atlanta schools. Since 2019, our team has recruited 27 STEM professionals and prepared them to become secondary science and mathematics teachers and teacher leaders. This body of work has explored (a) leveraging research-practice partnerships to support microteaching experiences (Louis & King, 2022), (b) Black science teachers as advocates and abolitionists in secondary science classrooms (Louis & King, 2023), and (c) sustainable self-care for early career STEM teachers (King et al., 2024).

Moreover, in my work with preservice teachers, I have collaborated with colleagues and doctoral students to disseminate evidence-based practices from methods courses in peer-reviewed and practitioner journals. This included (a) our use of modeling and computational thinking to understand waves (Enderle et al., 2021), (b) using genetic coding to learn about computational biology and STEM careers (King et al., 2018), and (c) the importance of surfacing students' prior knowledge in middle school science classrooms (Mesa et al., 2014).

In 2023, I was nominated and selected as a recipient of the National Science Foundation Alan T. Waterman Award—the nation's highest honor for early career scientists and engineers for my contributions to science education research and the STEM enterprise in advancing equity, fostering innovation, and cultivating the next generation of scholars. To date, I am the first (and only) educator to receive this recognition. The Waterman Award came with a medal and $1 million toward research initiatives. Dr. Alan T. Waterman was scientifically talented and accomplished in many regards, but most of all, he was a public servant. In honor of his great legacy and in alignment with my passions and scholarly endeavors, I am leveraging my current research agenda to further groundbreaking and transdisciplinary scholarship across the three aforementioned areas for which I was recognized. In 2023, I established the King Research Collaborative to serve as a hub for innovative research, community-engaged scholarship, and capacity-building efforts that center equity, justice, and humanizing practices in STEM education. This collaborative supports a growing network of scholars, educators, and community partners who are collectively reimagining what is possible in STEM teaching, learning, and research. Through the King Research Collaborative, we have collaborated with local partners to establish industry and community partnerships to provide real-world experiences and strengthen the STEM workforce. I also seek to expand and institutionalize pipeline programs where middle and high school girls can attend camps on the university campus and receive support to transition to college and careers through mentorship and scholarships.

Furthermore, we are expanding access to *I AM STEM* by increasing the number of partnerships with community-based organizations and children served. My team is creating a STEM learning ecosystem and providing access to resources and programs (in-person, hyflex, and online models) while helping schools and organizations build capacity to offer comprehensive and high-quality programs. The King Research

Collaborative is also offering opportunities for in-service teachers to receive professional development and support to become exemplary teachers and teacher leaders. We are creating an internal support system where experienced teachers can coach and mentor early-career teachers and working with the College of Education and Human Development to provide access to endorsements, certificates, and degrees within STEM Education to equip in-service teachers with the knowledge, tools, and resources to provide cutting-edge instruction to their students.

7.5 Turning a House into a Home through Teaching

Research indicates that Black women professors are subject to disproportionately negative teaching evaluations driven by racialized and gendered biases from students who perceive them as less competent or authoritative (Stanley, 2006). These evaluations, which frequently play a significant role in tenure decisions, often fail to reflect the innovative pedagogies and culturally responsive teaching practices that many Black women bring into classrooms. Thomas (2024) poignantly reflects on her experience in higher education, likening it to "bringing a folding chair" to a table that actively excludes Black women. This metaphor encapsulates the resilience and resourcefulness required of Black women who, despite systemic barriers, create spaces for themselves and others in the academy. In my own journey, I have been intentional about cultivating inclusive and affirming learning environments where students feel valued and empowered. Though I became a researcher through formal training, my passion for education began long before. I tutored throughout high school and landed my first teaching position for high school biology and chemistry at the age of 22.

Beyond the K-12 and out-of-school time sectors, I find great fulfillment in designing, (re)developing, and teaching courses at the university level. My seminars and methods courses equip students to become highly effective educators, practitioners, and researchers who are not only knowledgeable but also compassionate and culturally competent. I think critically about the types of experiences that I afford, and partner with local schools and organizations/programs to bring this vision to fruition. I spearheaded the development of a 100% online 36-credit hour masters degree program in STEM Education with four concentration areas and an embedded STEM Endorsement. I also created an 18-credit hour graduate certificate program which is now approved through the University System of Georgia's Board of Regents. The transdisciplinary certificate and degree programs are cutting-edge, engaging faculty across colleges and departments, and preparing science and mathematics teachers to humanize their curriculum and pedagogical practices by explicitly focusing on justice and equity. Beyond my role as a coordinator for the master's certificate and degree programs, I mentor postdoctoral research associates, serve as a major advisor or committee member for doctoral students, advise master's level students, and supervise undergraduate students on their research projects.

Recognizing that STEM careers are growing at a faster pace than any other job segment in our national economy, I have established relationships with industry partners such as Microsoft to provide real-life applications, upskill teachers, and expose them to new and exciting careers. I partner with Microsoft Philanthropies (Microsoft Skills for Social Impact) and World Wide Technology (WWT) to deliver scalable Artificial Intelligence (AI) and cybersecurity training to communities globally, taking a multi-generational approach, from youth to seniors to ensure broad access to these skills. Through this collaboration, we have an extensive plan for both virtual and in-person workshops with I AM STEM and Georgia State University. The workshops focus on preparing educators and youth professionals who, in turn, teach students across elementary, middle, high school, and college levels. Through this model, we expect to reach over 7,000 people. Consequently, the King Research Collaborative is producing 21st Century STEM Educators who are equipped with scientific, mathematical, and digital literacies to teach children in schools.

7.6 Turning a House into a Home through Service

While I am deeply committed to research and teaching, it is service that centers, grounds and informs my life's work. My love for children and the community drives me to engage in meaningful scholarship, publish in top-tier journals, secure grant funding, and provide me with creative energy to develop courses on intersectionality in STEM while also designing degree programs that engage faculty across colleges and disciplines for a shared purpose and goal. I alluded to the community-based work and hope to shed light on my engagement in and connection with the science education community. I actively contribute to the field through leadership and service in several professional organizations. I have previously served as Chair of the American Educational Research Association's (AERA) Out-of-School Time Special Interest Group (SIG) and the Multicultural Education SIG for the Association of Teacher Educators (ATE). Additionally, I was a strand coordinator for Cultural, Social, and Gender Issues within the National Association for Research in Science Teaching (NARST) and chaired the NARST Research Committee. Currently, I serve as President of the Southeastern Association for Science Teacher Education (SASTE) and as a member of the Elections Committee for the Association for Science Teacher Education (ASTE). In addition to my organizational leadership experience, I have also served as an invited reviewer for several academic journals and advisory boards for federally funded grant projects. I also serve on the committee for *Advancing AI in Science Education (AASE): A Comprehensive Approach to Equity, Inclusion, and Three-Dimensional Learning AI4STEM*. In editorial roles, I was a section editor for *Science Education*, an associate editor for the *Journal of Science Teacher Education (JSTE),* and I continue to serve as an editor for the *Journal of African American Women and Girls in Education.* My expertise extends to research funding and evaluation, including reviewing grant proposals for the National Science Foundation (NSF) and serving on expert review panels for the National Aeronautics and Space

Administration's (NASA's) STEM Engagement Performance Assessment and Evaluation. I take great pride in contributing to the field not only through cutting-edge research but also by mentoring and supporting emerging scholars. Beyond professional organizations, my service extends into the community through partnerships with institutions of higher education, informal STEM and faith-based organizations, industry, and K-12 schools. These collaborations reinforce my commitment to bridging research, practice, and community engagement to advance equity and access in STEM education.

7.7 Navigating Hardships at Home

While I have experienced success in my academic career as a Black woman science educator, my journey has not been easy. The plight of Black women in the academy is marked by a complex web of systemic challenges and invisible labor that underscores the intersection of race, gender, and institutional inequities. Despite significant contributions to scholarship, teaching, and service, Black women professors face persistent marginalization within higher education. As McGee and colleagues (2023) note, the academy often replicates societal inequities, placing the burden of "racial activism" and institutional transformation disproportionately on Black women. This expectation, coupled with the under-recognition of our contributions, positions Black women as both the architects of change and the victims of institutional neglect, perpetuating cycles of overwork and undervaluation.

Black women are often tasked with invisible labor that extends far beyond formal job descriptions, such as mentoring Students of Color, serving on diversity committees, and navigating the emotional toll of combating systemic racism in academic spaces. This work, although critical, is rarely accounted for in tenure and promotion processes, leaving Black women overextended and underappreciated (Griffin & Reddick, 2011; Harley, 2008). As McGee (2024) highlights, these demands create a paradox where Black women must "die to succeed," with their physical and emotional well-being sacrificed for the sake of professional advancement. The unrelenting pressure to work "twice as hard" persists even as the fruits of these efforts remain overlooked, reinforcing a double bind that is both exhausting and demoralizing.

Several years ago, I wrote an editorial on becoming Black girls and women in STEM education and careers (King, 2021). I noted that being *the first* comes with many challenges and moments of insecurity. The stakes are high as you shoulder the weight of unseen history and struggle to fit in while still standing out. There is often an untold story of struggle, resistance, and resilience; one that often fuels a deep commitment to outreach and advocacy.

In addition to being the first educator to receive the Alan T. Waterman Award, that same year, I was the first African American woman to receive the National Association for Research in Science Teaching's (NARST) Early Career Research Award—and only second African American (Dr. Bryan Brown at Stanford University was the first). I was not the first African American woman to be nominated, but the first to be recognized. I hope to serve as an inspiration for Black and Brown scholars who have felt that their work was overlooked or told that their scholarship is "too narrow" or "not important". I also challenge our colleagues who serve on these committees to acknowledge the work of groups who have been underrepresented and know that this is not a level playing field. I may never get the number of publications or citations of those connected to lineage within science education, but that does not make my contributions less important. We must also acknowledge that some scholars are at institutions where they have heavy teaching loads, invisible labor, or infrastructures that make it difficult to secure extramural funding. So, what is impact, who gets to decide it, and are we willing to think differently about what it means for our field moving forward?

As we consider the role of mentorship, we should seek to go beyond traditional mentorship to sponsorship (King & Upadhyay, 2022). This requires leveraging our positions of influence to advocate on behalf of mentees and provide endless opportunities. This form of mentorship is particularly important for Women of Color in the academy. As trusted scholars in the academy, you see and validate rising scholars in the field. Like mentorship, community is essential for not only surviving but thriving. My amazing colleagues, students, official and unofficial mentees, as well as brothers and sisters in the science education community hold me up and hold me accountable. There are many who support me publicly and those who have supported me in quiet ways (by reviewing and providing feedback on my manuscripts, writing external letters for promotion and tenure, recommending me for awards and recognitions, and serving as program officers and reviewers for external grant funding). Their investment in my life is immeasurable and has not gone unnoticed.

Dr. Kerry Ann Rockquemore, founder of the National Center for Faculty Development and Diversity, is the lead author of the book titled, "The Black Academic's Guide to Winning Tenure without losing your soul" (2008). One of my greatest accomplishments was earning tenure on my terms by doing what mattered to me. Black women are often evaluated against standards that fail to account for the intersectional impact of race and gender on their scholarly trajectories. As Harley (2008) and Griffin and Reddick (2011) argue, the metrics of success in academia are rooted in Eurocentric norms that privilege publication quantity over community impact, disproportionately disadvantaging Black women scholars who prioritize engaged and transformative scholarship. McGee et al. (2023) advocate for "equity ethics" to be institutionalized in STEM and higher education, calling for a radical reimagining of evaluation criteria to recognize the unique contributions of Black women and foster equitable academic environments. Black women faculty also navigate hostile environments where microaggressions and overt discrimination are routine, further eroding their sense of belonging and career satisfaction. The cumulative impact of these experiences is a high rate of burnout and attrition among Black women in academia, as they face systemic barriers to success without adequate institutional support. Efforts to address these inequities must prioritize structural changes that redistribute labor equitably, dismantle biases in evaluation processes, and foster inclusive cultures that genuinely value the contributions of Black women scholars (Tuitt et al., 2009).

7.8 Finding Our Way Back Home: A Purpose-Driven Approach to Tenure

We must engage in meaningful work that fills our spirit and not just our CVs. I call this a "convictions-based" approach because it has guided who I am, who I am becoming, and how I choose to show up in the academy. I constantly remind myself that I did not get here alone. Therefore, I refuse to be isolated and instead invest my energy in building bridges and finding my people. This is a lesson for early career scholars and those on the tenure track: remain steadfast in the face of resistance. I draw strength from knowing that my work is divinely inspired. Even in moments of opposition and doubt, the outcomes reaffirm the divine origin of my mission. I have stood firm, and my confidence in God and His purpose has never wavered.

For tenured and senior faculty, do not be so guided by theory that you neglect praxis. Step into the community and reposition yourself as a learner. You do not have to be an expert in every space you occupy. It is okay to sit at the feet of those in the communities you intend to serve. In that position, figuratively wash their feet. Seek to serve rather than be served. To borrow another biblical reference, do not betray the people you claim to serve for thirty pieces of silver...whether that be federal grants, tenure, or personal ambition. Do not shackle the same feet you once washed and fail to credit those who have provided the wisdom and insight to do this work. Trusting, reciprocal, and enduring relationships are what we need.

For those in administrative positions, remember what it felt like to be faculty. Use your power and privilege for good, not for personal gain or upward mobility. Challenge systems and institutional structures that oppress the most marginalized, especially if you do not belong to those groups or directly benefit from taking that stance. We need co-conspirators who are willing to risk their own privilege to dismantle the systems, structures, institutions, policies, and practices that perpetuate inequity.

For those of us in industry or leading our own businesses, may capitalism and corporate greed never define our purpose, decisions, or initiatives. In this current political climate, where economic and social disparities continue to widen, we must be intentional in resisting exploitation and centering equity in our work.

For those of us in classrooms, children are not blank slates. They are whole human beings, designed to be loved and nurtured, not simply managed. Our goal should not be to seek visibility for ourselves but to create pathways for others to be seen. Not always to be heard, but to listen intently. Not always to be understood, but to seek understanding of those whose realities and lived experiences differ from our own.

If you can accomplish your life's work solely in your own strength, then it is not big enough. Transformative work requires us to transcend self-interest and to walk in solidarity and community with others...without "othering."

7.9 Bringing Impact Home: Reimagining Success in the Academy

What is impact? Who gets to define it, and who determines how it is measured? More importantly, what is the real impact we hope to make in the academy and beyond? How will we collectively redefine what impact means in our field moving forward?

We have the power to reject traditional mechanisms that attempt to quantify our value. We can problematize and dismantle the current metrics of success and reimagine what it means to have a true impact. I have resisted conforming to a system that was never designed with me in mind, and that resistance has been both liberating and life-changing. Let us build something new and create a framework that gives scholars purpose and hope. A system that promotes collaboration and community over individual achievement.

During the Civil Rights Movement, Malcolm X reminded us that if we do not stand for something, we will fall for anything. I stand for and with the communities I serve. My laboratory is in the communityfully equipped for scientific investigation, research, teaching, and service. I have learned that you do not have to be an uptight, socially awkward, old white man to be a scientist or a science education researcher. You can be energetic, bold, and an extrovert. My life's experiences have taught me that you can also be a young, gifted, and unapologetically Black woman.

7.10 Finding My Song at Home

So many ask, "Dr. King, how do you do it?" Well, it's never about the "I" but the collective "we." It's the people who believe in me and push me every single day. It is my conscious and intentional decision to boldly show up as myself in each space that I occupy. It is realizing that while I am only a millennial, I have become an auntie and elder, in some cases setting a new precedent for what it means to "do" this thing called academia. It is the sum of my daily actions. Not who I pretend to be at a conference in the spotlight or in a peer-reviewed journal, but who I choose to be in a faculty meeting, in my research lab with students, or even at home with my family. It is who I am around those with whom I am comfortable and familiar—when I think that no one is looking. Walk in integrity. Show up for people and expect nothing in return but for them to pay it forward. Mahalia Jackson sang, "If I can help somebody, as I travel along. If I can help somebody with a word or a song. If I can help somebody from doing wrong. No, my living shall not be in vain." So never lose your song, soul, and sense of purpose.

I was an undergraduate pre-med student when I lost my song and felt like I almost lost my way. Slowly, I am finding my voice again. My song may not make it to the Top 20 MTV charts or win a Grammy, but it is mine. On days when the work gets hard, remember your song and keep singing. Sing like no one is looking. Sing like your voice is reverberating off hard, smooth surfaces in the shower, creating beautiful notes you did not think you had. Sing with a sense of urgency and power like Mahalia Jackson, or negro spirituals filled with hope like Harriet Tubman. Sing with a deep baritone voice that can soothe the soul like Luther Vandross or belt high notes that may earn you a duet with Mariah Carey (in the 1990s and early 2000s).

Speak so that people will listen. Sing so that people will feel. Dr. Yolanda Sealey-Ruiz encourages us to love from the vortex. Love that is not lip service only but a result of doing and being (2020). In the words of Bob Marley, "won't you help to sing these songs of freedom? Cause all I ever have are redemption songs." May the frequencies of our song vibrate and penetrate our life's work so that after we are gone, the true impact of our presence in this space will endure.

7.11 When God Calls Me Home

The academy was my first professional home after graduation, but it is not my final destination. While I am young, I realize that life is fleeting, temporal, and we are all just passing through. This world is not my home, but while I am here, I want to leave it better than how I found it when I arrived. I opened the chapter referring to a scripture in Acts 5:27–39 where the apostles had been arrested and brought before the Sanhedrin for defying orders to stop teaching in Jesus' name. After hearing their bold defense, the council was furious and wanted to kill them. It was at this point that Gamaliel stood up and offered his counsel. Gamaliel's point was clear: human-originated movements fade away on their own. However, if the apostles' mission was truly from God, no amount of human effort could stop it. Attempting to do so, he warned, would mean fighting against God Himself. Gamaliel's wise counsel to leave the apostles alone resonates with my own journey as a Black woman science educator. My work in advancing access and equity in STEM education is not merely a career choice but a divine assignment. Fighting against what God purposed for me to do and His call on my life are likened unto fighting against God Himself. He has ordained me to be a beacon of light in dark places and to shine for the world to see. Not me, but Christ who lives in and through me.

I have an unwavering commitment to a divine mission. Recounting the events of Acts 5, we see the apostles' steadfast dedication to their calling despite opposition. Their work disrupted the status quo, yet their purpose, divinely ordained, could not be stopped by human resistance. There are clear parallels to my journey. My personal convictions and role in the academy are not arbitrary; they are divinely orchestrated. My work challenges systemic inequities in science education disrupting exclusionary practices, advocating for curricular practices that represent and embrace the diversity of our children, mentoring doctoral students, and ensuring that our local and global communities benefit from research that is actionable and just. I recognize that the impact of my work is undeniable, and there is evidence of God's hands on every project and initiative to fulfill a much higher purpose.

This journey has been far from easy. Doing transformative work requires patience and discernment rather than impulsive action. It demands faith in God's sovereignty, knowing that human plans cannot thwart divine will. Like the apostles, I have come to understand that when a mission is rooted in God's purpose, it will persist despite obstacles. This conviction has guided me in aligning my expertise with a higher calling by formulating new theories, developing curricula, providing professional development for teachers, and mentoring future scholars. These contributions are not just professional duties; they are a sacred responsibility.

As a science educator, I have found my home not in the traditional metrics of academia but in the impact of my work on the lives of those I serve. Rejecting the scientometric index and conventional measures of success has been both liberating and life changing. I have chosen to problematize and disrupt existing notions of impact, redefining them through service, advocacy, and transformation. The community is where I find refuge; the community is my laboratory; the community is where I call homeuntil God calls me home.

References

Bell, D. (1992). *Faces at the bottom of the well: The permanence of racism.* Basic Books.

Brown, J. S., Collins, A., & Duguid, P. (1989). Situated cognition and the culture of learning. *Educational Researcher, 18*(1), 32–42.

Clandinin, D. J., & Connelly, F. M. (2000). *Narrative inquiry: Experience and story in qualitative research.* Jossey-Bass.

Creswell, J. W., & Creswell, J. D. (2017). *Research design: Qualitative, quantitative, and mixed methods approaches.* Sage publications.

Delgado, R., & Stefancic, J. (2017). *Critical race theory: An introduction.* NYU Press.

Emdin, C. (2016). *For white folks who teach in the hood... and the rest of y'all too: Reality pedagogy and urban education.* Beacon Press.

Enderle, P., King, N. S., & Margulieux, L. (2021). What's in a wave?: Using modeling and computational thinking to enhance students' understanding of waves. *The Science Teacher.*

Flick, U. (2007). *Managing quality in qualitative research.* Sage.

Gillborn, D. (2005). Education policy as an act of white supremacy: Whiteness, critical race theory, and education reform. *Journal of Education Policy, 20*(4), 485–505.

Gramsci, A. (1971). *Selections from the prison notebooks.* International Publishers.

Griffin, K. A., & Reddick, R. J. (2011). Surveillance and sacrifice: Gender differences in the mentoring patterns of Black professors at predominantly White research universities. *American Educational Research Journal, 48*(5), 1032–1057.

Harley, D. A. (2008). Maids of academe: African American women faculty at predominantly White institutions. *Journal of African American Studies, 12*(1), 19–36.

hooks, b. (1990). *Yearning: Race, gender, and cultural politics.* South End Press.

hooks, b. (1994) *Teaching to transgress: Education as the practice of freedom.* Routledge.

King, N. S. (2017). When teachers get it right: Voices of Black girls' informal STEM learning experiences. *Journal of Multicultural Affairs, 2*(1), 5.

King, N. S. (2019). *Let the church say amen to STEM: Guidebook to launching and growing extraordinary youth programs.* I AM STEM, LLC.

King, N. S. (2021). Toward an equity agenda for Black girls and women in STEM learning spaces and careers: Noticing, validating, and humanizing. *Journal of African American Women and Girls in Education.* https://doi.org/10.21423/jaawge-v1i2a93

King, N. S., Collier, Z., Johnson, B., Acosta, M., & Southwell, C. (2021). Determinants of Black families' access to a community-based STEM program: A latent class analysis. *Science Education.*

King, N. S., Jett, C. C., & Dziengue, A. (2024). *"Sharpen the saw": Sustainable self-care for early career STEM teachers.* AAAS Advancing Research & Innovation in the STEM Education of Preservice Teachers in High-Need School Districts (ARISE) Blog. https://aaas-arise.org/2024/04/11/sharpen-the-saw-sustainable-self-care-for-early-career-stem-teachers/

King, N. S., & Pringle, R. M. (2019). Black girls speak STEM: Counterstories of informal and formal learning experiences. *Journal of Research in Science Teaching,* 539–569.

King, N. S., & Upadhyay, B. (2022). Negotiating mentoring relationships and support for Black and Brown early-career faculty. *Science Education, 106*(5), 1149–1171.

King, N. S., Wade-Jaimes, K., & Morgan, P. D. (2018). Decoding careers in DNA. *The Science Teacher, 85*(5), 54–59.

Ladson-Billings, G. (1989). A tale of two teachers: Black teachers, culturally relevant pedagogy, and the problem of African American achievement. *Educational Researcher, 18*(2), 3–12.

Ladson-Billings, G. (1995). Toward a theory of culturally relevant pedagogy. *American Educational Research Journal, 32*(3), 465–491.

Louis, V. N., & King, N. S. (2022). Emancipating STEM education through abolitionist teaching: A research-practice partnership to support virtual microteaching experiences. *Journal of Science Teacher Education, 33*(2), 206–226. https://doi.org/10.1080/1046560X.2021.2012957

Louis, V. N., & King, N. S. (2023). "Each day, I find ways to fight for my students": Black science teachers as advocates and abolitionists. *Equity & Excellence in Education, 56*(4), 608–621.

Mesa, J. C., Pringle, R. M., & King, N. (2014). Surfacing students' prior knowledge in middle school science classrooms: Exception or the rule?. *Middle Grades Research Journal, 9*(3).

McGee, E. O. (2020). *Black, brown, bruised: How racialized STEM education stifles innovation.* Harvard Education Press.

McGee, E. O. (2024). Dying to succeed: Unveiling the (un)hidden toll of academic advancement for Black women. *Qualitative Inquiry.* https://doi.org/10.1177/10778004241269969.

McGee, E. O., & Bentley, L. (2017). The troubled success of Black Women in STEM. *Cognition and Instruction, 35*(4), 265–289.

McGee, E. O., Morton, T. R., White, D. T., & Frierson, W. (2023). Accelerating racial activism in STEM higher education by institutionalizing equity ethics. *Teachers College Record, 125*(9), 108–139.

McIntosh, P. (1988). White privilege: Unpacking the invisible knapsack. *INdependent School, 49*(2), 31–36.

Rockquemore, K., & Laszloffy, T. A. (2008). *The black academic's guide to winning tenure—Without losing your soul* (p. 261). Lynne Rienner Publishers.

Sealey-Ruiz, Y. (2020). *Love from the vortex & other poems.* Kaleidoscope Vibrations, LLC.

Stanley, C. A. (2006). Coloring the academic landscape: Faculty of color breaking the silence in predominantly White colleges and universities. *American Educational Research Journal, 43*(4), 701–736.

Thomas, V. G. (2024). If they don't give you a seat at the table, bring a folding chair: My HBCU faculty experiences. In *Journeys of Black Women in Academe: Shared Lessons, Experiences, and Insights* (pp. 103–120). Emerald Publishing Limited.

Tuck, E., & Yang, K. W. (2012). Decolonization is not a metaphor. *Decolonization: Indigeneity, Education & Society, 1*(1), 1–40.

Tuitt, F., Hanna, M., Martinez, L., Salazar, M., & Griffin, R. (2009). Teaching in the line of fire: Faculty of color in the academy. *Thought & Action, 25,* 65–74.

Chapter 8
Fitting in: A Black Woman's Perception

Rhea Miles

8.1 Introduction

A good challenging puzzle includes a few to as many as thousands of pieces to complete a desired picture. People spend days, weeks, months, or years to complete a puzzle either by themselves or with the assistance of others. On some occasions, there are puzzle pieces missing or pieces that do not fit anywhere. In many ways, parts of my life have been a living puzzle. However, my pieces were not pre-cut, ill-designed, or missing. My beautiful puzzle is composed of a sense of belonging, persistence, identity, and resilience, which are all molded by me and other people who helped me create a brilliant and complex picture of the scientist and science educator trying to fit in, Rhea Miles.

8.2 Sense of Belonging in STEM

Trujillo and Tanner (2014) evaluated several studies to measure a sense of belonging, particularly for persons working in fields related to science, technology, engineering, and mathematics (STEM). Their research concluded that a sense of belonging was a feeling of support and fitting into a physical space where an individual feels comfortable and respected. Additionally, Good et al. (2012) identified factors such as feeling valued and accepted as important to having a sense of belonging in professions related to mathematics. Furthermore, Rainey et al. (2018) defined a sense of belonging for scientists to include the following components: interpersonal relationships, STEM identity, personal interests, and competence. All these factors and components relate

R. Miles (✉)
East Carolina University (ECU), Greenville, NC, USA
e-mail: milesr@ecu.edu

M. M. Atwater and B. A. Garvin (eds.), *Creating and Maintaining a Home in Science Education Research*, Contemporary Trends and Issues in Science Education 65,
https://doi.org/10.1007/978-3-031-99382-4_8

to "fitting in" to science-related fields and contribute to the complexity of creating a person that has no "holes" or missing pieces in their sense of belonging.

8.2.1 Sense of Belonging for Black Women in a STEM Family

However, a sense of belonging is not the same for all populations in STEM. Specifically, for women and underrepresented groups in the United States, environmental interventions are key to developing a sense of belonging (Rainey et al., 2018). For example, I had a mother and father who both had degrees in mathematics (picture of my father and me seen in Fig. 8.1). My mother's first job was as a mathematics teacher, and my father was a statistician for a telephone company (Miles, 2019a). As described in Morton and Parsons (2018) research, having family members in STEM can be inspirational and motivational for Black female majors in STEM. It helped me "fit in" and create a sense of belonging for me in my chosen college major. In addition to my father's and mother's influence on my sense of belonging, my father's brothers had degrees in STEM, one as an aerospace engineer and the other as a computer software engineer (Miles, 2019a). My parents and uncles encouraged my siblings and me to participate in as many mathematics and science activities outside of school during our elementary, middle, and high school education.

My sister and I showed an intense interest in STEM and participated in a variety of summer STEM enrichment programs (Gordon, 2022). These experiences provided further environmental interventions that influenced a sense of belonging. Participating in informal science and mathematics enrichment programs nurtured my interests in STEM. These experiences allowed me to have hands-on experiences, preparing me for my future career. According to Ireland et al. (2018), Black females do not have these opportunities and feel underprepared for a future career in STEM. However, this was not the case for me. I have fond memories of conducting scientific investigations, discovering nature, completing scientific tasks, developing scientific skills,

Fig. 8.1 Picture of my father and me

and working as an individual and in groups to solve STEM-related problems because of my participation in STEM informal programs (Gordon, 2022).

8.2.2 Sense of Belonging and Dress for Black Women

Hunter et al. (2019) also found that the Black females interviewed described acculturative loss of their Black culture. This is described as a sacrifice of giving up culture to "fit in." I was questioned by my colleagues as to why I "dressed up" for work during the day at the university, which I describe as business attire. My family instilled in me that you always dress in your best clothes when giving a presentation or teaching. My late grandmothers and other relatives who have passed would be "rolling over in their graves" if they ever saw me wearing jeans to work. Accultural loss was NOT applicable to me.

Dress is associated with cultural connection with and responsibility to others, and in my case, it connected me to honoring my family. Hunter et al. (2019) further discuss how Black students enrolled in an introductory psychology course did not want to perpetuate a negative stereotype and desired to promote a positive image via clothing choices. Furthermore, McCluney and Rabelo (2019) described an incident when a Black female professional arrived at work, and a White female directed her to join the wait staff, even though she was not a waitress. This assumed reaction to her presence does not create a "sense of belonging." I was mistaken for a custodian when I arrived at the university where I work. This assumption about my presence at the workplace adds to the feeling of being invisible. McCluney and Rabelo (2019) describe the White majority workplace personnel as seeing Black females in certain subservient positions as the norm. To this day, I am proud to have relatives who were and are custodians, as they are intricate parts of the puzzle pieces which define me. Every person wants to be seen, including Black female STEM professionals who may seek out opportunities to be as visible and unique as possible by not conforming to fit in by choosing unique hairstyles or dress.

Furthermore, perhaps my actions of "dressing up" at work with my White colleagues are founded in my family's perception of not wanting to risk confirming a negative stereotype. Steele and Aronson (1995) discuss this as a "stereotype threat". It was instilled in me not to be thought of as beneath or of a lesser social class than White people. This is also a part of "Double Consciousness" defined by W.E.B. DuBois (Study.com, 2024) where in the Black Culture and equality and dignity are believed and encouraged, but at the same time, Black people have to realize that some White people perceive all Black people in a lower social status and as having lesser intelligence.

Additionally, part of finding a sense of belonging involves having allies without assimilating to the dominant male work culture, which may equate to not wearing business attire. According to the National Science Board's Science and Engineering Indicators (2024), men account for 65% of ALL workers in STEM. The masculine-dominated work environments can deter women from pursuing a career in STEM.

Seventy-two women out of ninety-six participants in a research study of computer science majors were dissuaded from working at a computer science company because they saw stereotypical male objects on display (Cheryan et al., 2009).

The percentage of Black males and females are under- represented also in the STEM workforce (National Science Board, 2024). According to Ross et al. (2021), Black women, specifically in engineering, can be made to feel out of place and not belonging or fitting into the norm for their chosen career. Ross et al. (2021) report a revelation from Black females' perspective in STEM of being comfortable with being the "only" in the workspace, which has led to confidence about working alone or dressing differently. The only in this context means being the one female or the one Black person of color, or both in the workspace, and, in turn, according to research, this confidence is an asset to and trait of good leadership, which I hope I, too, have illustrated.

8.3 Persistence in STEM

While a sense of belonging is a key factor in a successful path for a career in STEM, it is not mutually exclusive from persistence. Persistence is also like a single puzzle piece that cannot represent, nor can it show the entire picture of a professional in STEM without the other pieces (i.e., sense of belonging). Human persistence means a person will "stick" to what he or she intends to do. Persistence is a vital element needed to continue in a STEM-related field despite obstacles, such as those faced by women such as gender discrimination or limited interpersonal support (Wilkins-Yel et al., 2022).

8.3.1 Persistence of and Microaggression Toward Black Women in STEM

Persistence is a common theme for women and persons of color in science-related careers. In the article "Cell Press," published in the journal *Molecular Cell*, the documented personal experiences of scientists from underrepresented groups, such as Black women, were similar to mine (McGill et al., 2021). Despite my experiencing microaggressions, such as "Well, at least we know you are competent," I persisted. Moreover, the authors also report on how Black STEM female professionals felt uncomfortable when asked, "Is that your hair?" and were not acknowledged during professional meetings in group discussions. I, too, experienced these unacceptable comments and interactions but overcame them to continue my college studies and become a scientist and science education professor.

Microaggressions are perceived discriminatory aggressions, and according to Marshall et al. (2021), there are several types. Notably, the comment related to my competency as a professor in science education could be categorized as a microinsult. Another co-worker said to me in front of others about my best-selling children's book, "Oh, and it is really good." She appeared to be surprised that I, a tenured professor like herself, authored a popular book that was also well-written.

Unfortunately, from 1996 to 2016, the proportion of Black students with degrees in engineering decreased (Lee et al., 2020). Lee et al.'s (2020) research also noted that one possible reason for this decline could be microaggressions. Black students are perceived as inferior to having poor work ethic and low mathematics skills. Black females in their career paths in STEM have experienced racist remarks. I experienced an incident when a colleague loudly "called me out" for being late to a meeting, but when my White female counterpart was tardy or even absent, nothing was said.

Support systems include allies. Marshall et al. (2021) state that allies are colleagues or a peer support system at a place of work who bring attention to microaggressions. These persons advocate for victims of microaggressions to try to lessen the discrimination and de-escalate a hostile environment in the professional community. I do believe I have had allies where I work and appreciate the support in what I perceived as unfair situations.

8.3.2 *Persistence of and Support Systems for Black Women*

While Farias and Newheiser (2019) report that religious beliefs may not provide benefits from stress. However, to overcome obstacles during my career in science education, I have relied on my Christian faith to assist me in coping with microaggression so that I may persist in STEM. I pray to God when "things" are going well and when "things" are going not so well. I grew up in a religious family; my grandfather was a Baptist minister, and some family members believe in God while others praise Allah. I live by the biblical scripture "Faith without works is dead…" (*King James Online Version*, 2025, James 2:26). So, I worked hard as a scientist and science educator and believe that God knows my career path to success in STEM. Besides, I read that Galileo, a famous astronomer, was a religious person who faced opposition and persevered, so I believe I can too (Otunaya & Ago-Iwoy, 2020).

In addition to relying on my religious beliefs, my husband, whom I dated during my undergraduate work, also was and is a part of my support system, which I needed to overcome challenges in my life to achieve my STEM career goals. We both serve as instructors in higher education and have shared interests. Dyoga et al. (2022) concluded that married couples having higher education degrees, having been married for several years, and being established in each one's respective field were attributes needed to sustain a marriage during the Covid-19 pandemic. I believe these same qualities are important for each person in a marriage to support each other, whether there is or is not a pandemic.

I did not always feel supported attending classes as an undergraduate student. During my undergraduate work, I had one White male professor who, on more than one occasion, informed the entire class that women should not pursue a degree in chemistry. He stated this often while smiling, which caused others to laugh. Despite my professor's protestations, I persisted and earned an "A" in his class. According to a survey conducted by Ebrahimi et al. (2023), an unwelcoming environment due to gender bias is commonplace for women in biomechanics. However, it can be addressed at higher education institutions with mentorships and the formation of support groups to retain women in STEM careers.

Palmer et al. (2011) also report that persistence as a college STEM major is dependent on peer group support, involvement in STEM-related activities outside of class, and strong high school preparation in STEM-related coursework for students of color. Students of color who show interest in majoring in STEM, like me as an honor student in high school and having a passion for the subject matter in college, need to be encouraged to participate in STEM extracurricular programs to see how what is learned in the classroom is applied to a STEM career (Palmer et al., 2011). Ireland et al. (2018) research further acknowledged the importance of participation in summer research programs at the undergraduate level for Black women. Their research findings indicate that participation in pre-college STEM enrichment programs enhances the retention and graduation rate in college for Black female students. Ireland et al. (2018) also noted that participation in summer research programs positively affected career choices in STEM for Black women. As previously mentioned, I participated in informal learning programs from elementary through high school, and my interests and opportunities to participate in additional learning experiences in STEM continued in college. While an undergraduate student, I had the opportunity to conduct scientific research outside of my college classroom work and meet other chemistry majors with similar interests. The work I completed during undergraduate research one summer was later published in a peer-reviewed journal (Pharr et al., 1991). This experience, and my parents' pride attending a presentation related to this publication, made me feel like I fit in and influenced my decision to continue and persist as a chemistry major. I graduated from college with a degree in chemistry.

8.3.3 Persistence of Black Women in STEM in a Male-Dominated Culture

Black women have been able to navigate their studies for higher degrees at educational institutions dominated by a White male culture. The White male culture can negatively affect completing a STEM-related degree (Ireland et al., 2018). Ong et al. (2018) reported as well on Black students being perceived as not as smart or as intelligent as White males. Unfortunately, because of this misconception, working in a group or having a lab partner can be challenging at a White institution, so the women of color work in the lab alone, which I experienced. The authors also discussed being "the only," mentioned earlier, which can negatively affect persistence in STEM; however, it did not comparably affect me. It was my "normal" to work in isolation and to work without a lab partner or with a person thought to be incompetent in the chemistry laboratory (Lee et al., 2024).

After earning my Bachelor of Science degree in chemistry and a master's degree in chemistry education, I was employed as a chemistry teacher at the secondary school level. A few years later, I decided to pursue a doctoral degree. While seeking my PhD at a Predominantly White Institution (PWI), my doctoral defense committee member asked me why the doctoral committee should grant me a PhD. I was noticeably pregnant with my first child at the time of my defense, and I responded that I planned to *earn* my degree and seek employment as a professor at a four-year university. My friend, after telling her that the conversation bothered me, said that I should have responded, "I will hang my diploma on my wall and look at it." I did think to myself, would a man be asked this same question if his wife, girlfriend, or partner were expecting a child? Despite this questionable inquiry, I persisted and graduated with my doctoral degree.

Even when I was a doctoral student and an assistant professor expecting the birth of my second child, there were a few incidents where some of my White colleagues communicated to me that they assumed I was a single mother, which I was not. I questioned to myself whether this assumption would have been made if I were a White female. There is nothing wrong with being a single mom, but the way it was communicated to me was negative. These comments made me feel less than, like a stereotype, and unwelcomed. Ireland et al. (2018) state that throughout the educational and career pipeline for Black women in STEM, counseling resources should be made available to help them navigate this type of societal environment to persist in their STEM-related careers. I concur.

Black women who persist and remain successful (10 or more years of experience) in the engineering field learn how to navigate the White male-dominated field by reauthoring and redefining themselves to develop their identity (Ross et al., 2021). In addition, as described by Ross et al. (2021), despite barriers and challenges, Black women in STEM have persevered in their field of employment, in part from the support of family, friends, spouses, or colleagues at work. When I was invited as a keynote speaker, as an associate professor, at the university where I work (Miles, 2019b), not only did the local community from several civic organizations which I belong (such as the local chapters of Delta Sigma Theta Sorority Inc. chapter, Links Inc. Jack and Jill of America Inc. and National Council of Negro Women Inc.) attend my presentation, but so too did colleagues from my work at a PWI.

According to Ong et al. (2018), women of color do not pursue or are not represented in STEM at the same rate as their White male counterparts at places of work. Women of color who are persistent in STEM fields create counter-spaces; allowing people to see representatives of their respective group identities in STEM positions. Ong et al. (2018) discuss that counter spaces are necessary and important for women of color. However, since their persistence in STEM careers is often low, they do not often get opportunities to create these counterspaces. However, it can still happen in focused STEM environments. For example, I attended a conference focused on diversity, equity, and inclusion, and the speakers and leaders of the STEM session looked like me (STEM Awardees Meeting, 2022). As stated by Ong et al. (2018), having other people to relate to with similar experiences is important.

To increase persistence in STEM for women of color, a summary of Ong et al. (2018) suggestions are listed below:

1. Increase the visibility of gender and racial-ethnic diversity.
2. Seek partnerships with organizations serving women of color interested in STEM.
3. Organize informal talks and formal presentation workshops with women of color in STEM.
4. Create a department of anti-discrimination policies and consequences to guard against all forms of discrimination.
5. Sponsor department-wide meetings for students early each academic year, discourages, racial, ethnic, and gender discrimination and micro aggression.
6. Organize social, teambuilding events for faculty women of color.
7. Encourage faculty to prioritize reaching out and getting to know women of color.
8. Sponsor women of color students in any interest, faculties make presentations at STEM diversity conferences.
9. Create a campus chapter of STEM diverse organizations that is publicized and leads to other activities on campus as well as professional and alumni organizations.
10. Create a mentoring program for faculty and students.

The above suggestions effectively influence the persistence of Black women in STEM fields in higher education. Additionally, Morton and Parsons (2018) write about the need for support systems to be in place to have a presence of more persons of color to demonstrate that women of color, and Black people in general, can earn degrees in STEM, like White men, to debunk the persistent assumed negative stereotypes.

Researchers question the effectiveness of interventions to reduce devaluing cultures or changing prejudices (Paluck et al., 2021); However, Eagly and Koenig (2021) report that incremental change can affect stereotypes. An increase in female representation in the labor force from 1946 to 2018 resulted in a positive shift in the belief that women are just as intellectually competent as men. Moreover, over time, associating women with only artists, nurses, teachers, or family and men with science and careers has lessened. Similarly, Black women with social support working in non-traditional roles, such as careers in STEM fields, can affect stereotypes and persistence in STEM.

8.4 STEM Identity for Women

STEM identity, along with persistence and a sense of belonging, is also a vital part of pursuing a degree in STEM. STEM identity reflects how one sees oneself "fitting in" to the STEM profession. Singer et al. (2020) discussed how STEM identity is important to how an individual envisions himself or herself as a scientist, engineer, or mathematician. Key factors of environmental interventions that affect STEM identity include having a welcoming, supportive, and *non-perpetuation* attitude that the STEM-related workforce is for men only. According to Ramsey et al. (2013), STEM identity for women is impacted by environmental interventions, and Carlone and Johnson (2007) state that STEM identity is formed and influenced by recognition by society, which can be developed and formed during undergraduate and graduate studies.

8.4.1 STEM Identity for Black Women

As an undergraduate student, I was aware of only one female chemistry major who looked like me and graduated from my undergraduate university. I believed if she could graduate with a chemistry degree then so could I. This is also the basis of Gee's long-standing theory of identity, which is linked to the recognition of oneself by others (Gee, 2000). Representation mattered to me and other students. Black women must deepen their personal, familial, and cultural identity in STEM to further develop a STEM identity essential to success in the field (Ireland et al., 2018). Ireland et al. (2018) also report that key factors specific to influencing a Black woman's identity leading to success in a STEM profession include having a support system.

Ramsey et al.'s (2013) research indicated that having a female role model to identify with while pursuing a science degree in college assisted with developing STEM identity. Thus, Black female students in STEM need to have role models and instructors who look like them to identify with and positively influence them. I did not experience this, but research has shown that culturally responsive STEM exemplars (instructors and speakers) can cultivate Black STEM identity and perceptions of STEM at the elementary school grade level for ALL students (Hrabowski, 2018).

Olitsky (2020) further reports that finding Black females working in the STEM workspace is limited. While I did have female teachers in my elementary, middle, high school, and college career, I only had one Black male science high school teacher, I had no Black female science teachers during my K-12 years and college experience. It would have been nice to have a role model who looked like me to talk about my career path to inspire or motivate others to seek employment in a STEM-related field.

Sadly, statistically, many Black students have White teachers who have lower expectations of their Black and female students than White students (Tueresky et al., 2021). Black students may not be enrolled in advanced mathematics and science classes in middle and high school because they do not have role models to identify with in STEM, such as Black women in the STEM fields in education (Lee et al., 2024).

One reason the retention of Black females in science education is low may also be due to not having other Black females to self-identify as co-workers. Conflicts with identity in STEM for Black females affect recruitment and retention. Olitsky's

(2020) research reveals how this environment of identity conflict was challenging for a Black female science educator. I have worked with fellow Black female science educators as colleagues, which nurtured my STEM identity.

Once I became a Black female science educator, it was bolstering to be recognized for my efforts of excelling in the field. Similarly, Olitsky (2020) reports that while it was unexpected, a Black female teacher in a research study was acknowledged and appreciated for work well done. This positively affected this teacher's STEM identity which was not sought after but needed.

The percentage of Black women employed in STEM fields has remained extremely low (less than 5% of all persons in STEM in the United States) (National Science Foundation-National Science Board, 2024). Moreover, very few of these Black women have been recognized for their efforts. The *New York Times* best-selling book "*Hidden figures: The untold story of the African American women who helped win the space race.*" recognized the work of three Black women (who were "human computers") who endured gender and racial discrimination at the National Aeronautics and Space centers in the 1960s (Shetterly, 2016). This story inspired my children's book (see Fig. 8.2) about my father and uncles, who inspired me to pursue my current career and facilitated my identity in STEM.

Additionally, as Morton and Parsons (2018) further discuss, believing one can do science is part of the STEM identity. Black STEM identity is developed by the perception of resources and support from others. For example, Smith (2022) wrote that the family and community networks (i.e., church and Black Society of Black Engineers) provided the capital Black male engineering majors needed to succeed in college. Similar resources were available to me. Support from other people, such as financial familial resources, can influence STEM identity (Ireland et al., 2018). Most importantly, Collins (2018) further posits to foster a positive Black STEM identity, the Black family and community must value and appreciate STEM.

STEM should be associated with the cultural, social, and historical context of a Black woman's identity. Furthermore, the Black woman's emotional connection to STEM is important to forming her STEM identity (Avraamidou, 2020). She has to connect psychologically to others socially, seeking similar pursuits in STEM outside

Fig. 8.2 Picture of me with my children's book "*Adventures of the STEM Brothers*"

of the family (Ireland et al., 2018). In addition to the support of my family, I also had the opportunity to work in a study group of Black female chemistry majors pursuing master's and doctoral degrees, which assisted with nurturing and validating my identity in science. This type of learning experience was described as a shared learning experience resulting from having a support group, which also led to my positive trajectory in a STEM career.

During my undergraduate studies, I had a White male mentor who did not treat me differently or make me feel dissimilar like some of my other White male instructors did. He made me feel like a "real" chemistry major and recognized my work publicly with my peers and family. He told my parents in front of other college students that I had found a solution to a chemistry problem that he or others in the class had not considered. Avraamidou (2020) discusses how experiences like this developed my STEM identity. I enthusiastically pursued my degree and was compelled to work in science. My identity was what Avraamidou (2020) describes as being visibly nurtured and cared about as a Black female in science, which not only increased my self-esteem but also my self-efficacy in my desire to succeed in my STEM studies.

While pursuing a STEM degree and working in a STEM-related field, I realized and recognized that I identified with two oppressed social constructs—being female and Black (Ireland et al., 2018). These two intersectional realities can lead to feeling out of place, not belonging, or fitting in (Avraamidou, 2020; Ross et al., 2021). Avraamidou (2020) shared in her research how a Black female preservice teacher in college had opportunities to share intersections of ethnicity and gender led by a Black female instructor. She identified with class discussions that were relevant to her experiences, which made her feel valued. Ross et al. (2021) also suggested that Black female engineering majors must feel nurtured, supported, and feel a sense of belonging to fit in and develop their STEM identities.

In addition, Ireland et al. (2018) discussed, the intersectionality of being Black and female is not acknowledged. According to these authors, Black women's experiences with race, gender, and discrimination are overlooked, which results in feelings of being invisible. However, acknowledging negative experiences that affect identity can address positively the psychological disposition toward pursuing an education or career in STEM.

8.5 Resilience

Despite difficulties, a resilient person persists in overcoming adversity and continues to move forward in life despite horrible conditions (Rutter, 2006). Furthermore, a resilient individual wants to develop strategies and coping skills to not succumb to a defeatist attitude and remain positive and happy in unfortunate circumstances. A resilient person perseveres in his or her career and "bounces back" after rejection to achieve a goal that may seem impossible.

8.5.1 Resilience in STEM for a Black Woman

Black female engineers develop resiliency and identities during college, which can describe a Black female scientist and science educator like myself. Racism and sexism can be barriers to accomplishing goals; however, it is commonplace for Black women to navigate this experience of discrimination and still pursue a degree in STEM (Avraamidou, 2020; Morton & Parsons, 2018). According to Leyva (2021), Black women in STEM develop coping strategies to counteract negative stereotypes and discrimination. Furthermore, Black female STEM educators learn to oppose the racist and patriarchal structures in school systems despite being treated differently and facing inequalities (Ireland et al., 2018; Olitsky, 2020). These women become accustomed to others underestimating their abilities in STEM-related subjects such as mathematics or science and learn to be resilient. This means they can withstand difficult conditions and still complete tasks while serving in leadership roles.

Moreover, the Black female science educator must realize that there will be certain challenges that can be overcome in the workspace. Olitsky (2020) reported that a Black female science teacher's maternal approach to instruction included emotional and social connections to the Black and Latino students she taught. Her methodology opposed the patriarchal and high test-score-driven strategies of the school system culture. However, she was resilient and could negotiate her intersectionality to meet learning goals in her science classroom.

One coping strategy Black women use to overcome oppression in STEM is to form support as solidarity (Ireland et al., 2018). My sister and I were both undergraduate STEM majors; she in chemical engineering and I in chemistry, and we worked together on homework. This coping strategy assisted us with overcoming isolation while each of us pursued our degrees separately at two different PWI (Leyva, 2021).

I had become accustomed to a feeling of not being supported in middle and high school, where there were two other Black females in my honors and advanced placement classes who did not support my accomplishments. While in graduate school, White students would openly acknowledge that because I was a Black female, they assumed this fulfilled a quota related to "Affirmative Action"—still in effect when I was in college. Assumptions were made that I did not have the same qualifications as my White peers, which I did. These students seemed so nice. My mother often said, "Just because they are smiling at you or look like you does not mean they support you." I was resilient and continued to earn both my master's and doctoral degrees.

At the time of this chapter's development, some praise the end of "Affirmative Action" while others heavily scrutinize it. Although it has been suggested that the absence of "Affirmative Action" will be limited to only college or university admissions, the vacuum of its absence is also causing worrisome ideas. The university where I work more broadly interpreted the *Student for Fair Admissions, Inc. v. President & Fellows of Harvard College* ruling to apply to any program, which Totenberg (2023) projected would occur. My university's interpretation of the Students for Fair Admissions, Inc. v. President and Fellows of Harvard College (2023) extended to program for program efforts in a communication to me that "*Opportunities need to be open to all teachers. Demographic information should not be asked before enrollment but can be asked afterwards. Presenters should be chosen based on experience and have knowledge on the topic and not be chosen based on race nor ethnicity nor gender. Finally, there cannot be a "target" demographics of the participants."* My research interests are related to increasing the number of underrepresented and underserved groups interested in pursuing STEM-related careers in the United States. I am the principal investigator of a grant-proposed project to take place in school districts to increase interest in STEM for young children. Of course, the plan is to have these participants in the project to reflect the gender and ethnicity of student populations in the area where they live. University counsel perceived this as targeting the participant population which the project is not.

Additionally, I solicited engineers through several professional networks to serve as instructors and co-principal investigators for this project. The qualified self-selected engineers and engineering educators who agreed to work on this project were all Black. This caused the efforts to secure funding for ALL student participants to be scrutinized via the university counsel, which was unprecedented. Only through patience, persistence, and resilience was I able to get the grant project proposal submitted. The experience was challenging, but I am not deterred from seeking funding.

8.6 Conclusion

A sense of belonging, my STEM identity, persistence, and resilience have permitted me to recognize the puzzle pieces that fit together to create the picture of me in a STEM-related profession. For me, a sense of belonging is like being a part of a family STEM business. My mother, father, and uncles make me feel like I belonged as part of a STEM career legacy. Support systems, informal STEM experiences in elementary, middle, and high school, and participation in undergraduate research in college, provided me with what I needed to feel whole and nurtured my STEM identity. My persistence and resilience allowed me to matriculate as a chemistry major, earn a master's degree in chemistry education and a doctoral degree in science education, and persevere as a professor in science education despite challenging circumstances.

References

Avraamidou, L. (2020). Science identity as a landscape of becoming: Rethinking recognition and emotions through an intersectionality lens. *Cultural Studies of Science Education, 15*, 323–345. https://doi.org/10.1007/s11422-019-09954-7

Carlone, H. B., & Johnson, A. (2007). Understanding the science experiences of successful women of color: Science identity as an analytic lens. *Journal of Research in Science Teaching, 44*(8), 1187–1218.

Cheryan, S., Plaut, V. C., Davies, P. G., & Steele, C. M. (2009). Ambient belonging: How stereotypical cues impact gender participation in computer science. *Journal of Personality and Social Psychology, 97*(6), 1045–1060.

Collins, K. H. (2018). Confronting color-blind STEM talent development: Toward a contextual model for Black student STEM identity. *Journal of Advanced Academics, 29*(2), 143–168.

Dyoga, D. A., Herawati, T., & Defina. (2022). Quality of marriages of husband-wife families working in the formal sector during the Covid-19 pandemic as well as work-family conflicts and family interactions that occur. *Journal of Family Science, 7*(1). https://doi.org/10.29244/jfs.v7i1.39690

Ebrahimi, A., Daniels, K. A. J., Gaffney, B. M. M., Banks, C. L., McDonald, K. A., Kessler S. E., & Mahar J. N. (2023). International women in biomechanics: Promoting, supporting, and sustaining the careers of women in biomechanics. *Journal of Biomechanics, 146*. https://doi.org/10.1016/j.jbiomech.2022.111419

Eagly, A. H., & Koenig, A. M. (2021). The vicious cycle linking stereotypes and social roles. *Current Directions in Psychological Science, 30*(4), 343–350. https://doi.org/10.1177/09637214211013775

Farias, M., & Newheiser, A.-K. (2019). The effects of belief in God and science on acute stress. *Psychology of Consciousness: Theory, Research, and Practice.* https://doi.org/10.1037/cns0000185

Gee, J. P. (2000). Identity as an analytic lens for research in education chapter 3. *Review of Research in Education, 25*(1), 99–125. https://doi.org/10.3102/0091732X025001099

Good, C., Rattan, A., & Dweck, C. S. (2012). Why do women opt out? Sense of belonging and women's representation in mathematics. *Journal of Personality and Social Psychology, 102*(4), 700–717. https://doi.org/10.1037/a0026659

Gordon, J. (2022). Career imagination. In National Alliance of Black School Educators—NABSE. In *STEM century: It takes a village to raise a 21st century graduate* (pp. 77–82). CISTEMIC.

Hrabowski, F. A. (2018). The 2017 Charles H. Thompson lecture-colloquium presentation: Broadening participation in American higher education-a special focus on the underrepresentation of African Americans in STEM disciplines. *The Journal of Negro Education, 87*(2), 99–109. Retrieved from https://www.proquest.com/scholarly-journals/2017-charles-h-thompson-lecture-colloquium/docview/2091682854/se-2

Hunter, C. D., Case, A. D., & Harvey, S. I. (2019). Black college students' sense of belonging and racial identity. *International Journal of Inclusive Education, 23*(9), 950–966. https://doi.org/10.1080/13603116.2019.1602363

Ireland, D. T., Freeman, K. E., Winston-Proctor, C. E., Delaine K. D., Lowe, S. M., & Woodson, K. M. (2018). (2018). (Un) hidden figures: A synthesis of research examining the intersectional experiences of Black women and girls in STEM education. *Reviews in Research in Education, 42*(1), 226–254. https://doi.org/10.3102/0091732X1875907

King James Version online. (2025). https://www.thekingsbible.com/Bible/59/2?verse=26

Lee, A., Henderson, D. X., Corneille, M., Morton, T., Prince, K., Burnett, S., & Roberson, T. (2024). Lifting Black student voices to identify teaching practices that discourage and encourage STEM engagement: Why #Black teachers matter. *Urban Education, 59*(4), 979–1011. https://doi.org/10.1177/00420859211073898

Lee, M. J., Collins, J. D., Harwood, S. A., Mendenhall, R., & Huntt, M. B. (2020). "If you aren't White, Asian or Indian, you aren't an engineer": Racial microaggressions in STEM education. *International Journal of STEM Education, 7*(1). https://doi.org/10.1186/s40594-020-00241-4

Leyva, L. A. (2021). Black women's counter-stories of resilience and within-group tensions in the White, patriarchal space of mathematics education. *Journal for Research in Mathematics Education, 52*(2), 117–151. https://doi.org/10.5951/jresematheduc-2020-0027

Marshall, A., Pack, A. D., Owusu, S. A., Hultman, R., Drake, D., Rutaganira, F. U. N., Namwanje, M., Evans, C. S., Garza-Lopez, E., Lewis, S. C., Termini, C. M., AshShareef, S., Hicsasmaz, I., Taylor, B., McReynolds, M. R., Shuler, H., & Hinton, A. O. (2021). Responding and navigating

racialized microaggressions in STEM. *Pathogens and Disease, 79*(5). https://doi.org/10.1093/femspd/ftab027
Miles, R. (2019a). *Adventures of the STEM brothers*. TaylorMade Publishing.
Miles, R. (2019b). *Hidden no more: STEM women of color*. Office of Equity and Diversity East Carolina University.
McCluney, C. L., & Rabelo, V. C. (2019). Conditions of visibility: An intersectional examination of Black women's belongingness and distinctiveness at work. *Journal of Vocational Behavior, 113*, 143–152. https://doi.org/10.1016/j.jvb.2018.09.008
McGill, L., Brooks, D., Joseph, J., Guerrier, S., Addo, T., Guerra, A., Walker, M., Johnson, J., Ayombil, F., & Odinammadu, K. (2021). Voices on diversity: Creating a sense of belonging. *Molecular Cell, 81*(2), 213–217. https://doi.org/10.1016/j.molcel.2020.12.045
Morton, T. R., & Parsons, E. R. (2018). #BlackGirlMagic: The identity conceptualization of Black women in undergraduate STEM education. *Science Education, 102*, 1363–1393. https://doi.org/10.1002/SCE.21477
National Science Foundation-National Science Board. (2024). *Science and engineering indicators*. https://ncses.nsf.gov/pubs/nsb20243/key-takeaways
Olitsky, S. (2020). Teaching as emotional practice or exercise in measurement? School structures, identity conflict, and the retention of Black women science teachers. *Education and Urban Society, 52*(4), 590–618. https://doi.org/10.1177/0013124519873676
Ong, M., Smith, J. M., & Ko, L. T. (2018). Counterspaces for women of color in STEM higher education: Marginal and central spaces for persistence and success. *Journal of Research in Science Teaching, 55*(2), 206–245. https://doi.org/10.1002/tea.21417
Otunaya, A. O., & Ago-Iwoy. (2020). Religion and science: A short history of the existing skirmish and implications for Christianity today. *Pharos Journal of Theology, 101*. https://www.pharosjot.com/uploads/7/1/6/3/7163688/article_9_vol_101__2020__nigeria.pdf
Palmer, R. T., Maramba, D. C., & Dancy, T. E. (2011). A qualitative investigation of factors promoting the retention and persistence of students of color in STEM. *Journal of Negro Education 80*(4), 491–504. https://www.muse.jhu.edu/article/806879
Paluck, E. L., Porat, R., Clark, C. S., & Green, D. P. (2021). Prejudice reduction: Progress and challenges. *Annual Review of Psychology, 72*, 533–560.
Pharr, D., Selnau, H., Pickral, E., & Gordon, R. (1991). Enhancement effects of dodecyl sulphates in flame atomic absorption spectrometry. *Analyst, 116*(5), 511–515.
Rainey, K., Dancy, M., Mickelson, R., Stearns, E., & Moller, S. (2018). Race and gender differences in how sense of belonging influences decisions to major in STEM. *International Journal of STEM Education, 5*(10), https://doi.org/10.1186/s40594-018-0115-6
Ramsey, L. R., Betz, D. E., & Sekaquaptewa, D. (2013). The effects of an academic environment intervention on science identification among women in STEM. *Social Psychology of Education: An International Journal, 16*(3), 377–397. https://doi.org/10.1007/s11218-013-9218-6
Ross, M. S., Huff, J. L., & Godwin, A. (2021). Resilient engineering identity development critical to prolonged engagement of Black women in engineering. *Journal of Engineering Education, 110*(1), 92–113. https://doi.org/10.1002/jee.20374
Rutter, M. (2006). Implications of resilience concepts for scientific understanding. *Annals of the New York Academy of Sciences, 1094*(1), 1–12. https://doi.org/10.1196/annals.1376.002
Shetterly, M. L. (2016). *Hidden figures: The untold story of the African American women who helped win the space race*. Harper Collins.
Singer, A., Montgomery, G., & Schmoll, S. (2020). How to foster the formation of STEM identity: Studying diversity in an authentic learning environment. *International Journal of STEM Education*, 7(57), https://doi.org/10.1186/s40594-020-00254-z
Smith, D. T. (2022). They are here to support me: Community cultural wealth assets and precollege experiences of undergraduate Black men in engineering. *Journal of Engineering Education, 111*(4), 750–769. https://doi.org/10.1002/jee.20480

Steele, C. M., & Aronson, J. (1995). Stereotype threat and the intellectual test performance of African Americans. *Journal of Personality and Social Psychology, 69*(5), 797–811. https://doi.org/10.1037/0022-3514.69.5.797

Students for Fair Admissions, Inc. v. President and Fellows of Harvard College, 20-1199. (2023). https://www.supremecourt.gov/opinions/22pdf/20-1199_hgdj.pdf

Study.com. (2024). *W.E.B. Du Bois & double consciousness | definition & examples.* https://study.com/learn/lesson/video/web-du-bois-double-consciousness-overview-background-examples.html#:~:text=The%20internalization%20of%20such%20anti,current%20issues%20with%20double%20consciousness

STEM Awardees Meeting. (2022). *Burroughs welcome fund biennial stem education awardees conference,* Charlotte, NC. https://www.bwfund.org/news/2022-stem-awardees-meeting/

Totenberg, N. (2023, June). Supreme Court guts affirmative action, effectively ending race-conscious admissions. *National Public Radio–NPR.* https://www.npr.org/2023/06/29/1181138066/affirmative-action-supreme-court-decision

Trujillo, G., & Tanner, K. D. (2014). Considering the role of affect in learning: Monitoring students' self-efficacy, sense of belonging, and science identity. *CBE Life Sciences Education, 13*(1), 6–15. https://doi.org/10.1187/cbe.13-12-0241

Tueresky, K. M., Sinclair, S., Starck, J. G., & Shelton, N. J. (2021). Beyond, students: How teacher psychology shapes educational inequality. *Trends in Cognitive Sciences, 25*(8), 697–709.

Wilkins-Yel, K. G., Arnold, A., & Bekki, J. (2022). "I can't push off my own mental health": Chilly STEM climates, mental health, and STEM persistence among Black, Latina, and White graduate women. *Sex Roles, 86*, 208–232. https://doi.org/10.1007/s11199-021-01262-1

Chapter 9
At Home in the Interstices

Sheron L. Mark

In August 2023, I attended the International Sociology of Sport Association research conference in Ottawa, Canada. I presented research on framing science teaching and student engagement around sports and physical activity and problematized the engagement of race and culture in the contexts of sports and education. My presentation was titled, *Sociology of Sport Articulating with Teacher Education: Possibilities for Equitable and Culturally-Relevant Pedagogy.* This work was an extension of ongoing research published in a science/ Science, Technology, Engineering, and Mathematics (STEM) education edited book (Mark et al., 2022). My talk was well-received at this sociology research conference, and I believe that it fit seamlessly, as my scholarship and teaching have been situated at the intersections of critical race studies, the sociology of race, education and culture, and teacher education. A satisfactory and fulfilling home in academia for me had to be one that was intersectional and situated in and beyond science education.

Finding this home has been a circuitous and ongoing journey. Following undergraduate studies in biochemistry, I pursued graduate studies in biomedical and chemical engineering; but I did not feel settled. Attention to and the study of power and culture were missing. I was interested in issues pertaining to education and young people. I also wanted to have an impact on teachers and university faculty. As I prepared to apply for doctoral programs upon my Master's graduation, in addition to curriculum and instruction, I considered programs in women and gender studies, sociology, anthropology and psychology. Ultimately, I would complete doctoral studies in curriculum and instruction with a focus in science education before responding to that still unsettled calling in sociology.

So, from very early on, I always looked widely across disciplines for my place in academia; but, importantly, these all fit together meaningfully from my perspective.

S. L. Mark (✉)
College of Education and Human Development, University of Louisville, Louisville, KY, USA
e-mail: sheron.mark@louisville.edu

M. M. Atwater and B. A. Garvin (eds.), *Creating and Maintaining a Home in Science Education Research*, Contemporary Trends and Issues in Science Education 65,
https://doi.org/10.1007/978-3-031-99382-4_9

The diverse array of disciplines and fields of study were never, to me, "either/or" decisions, but rather "and" opportunities. These were exciting and intellectually invigorating to me. Possibilities seemed endless. Boredom seemed less likely as I can continue to grow as an academic, teaching and conducting research in various arenas and within diverse collaborations. At the same time, there have been drawbacks. As I proceeded through my studies in education and, later on, into my faculty position as an Assistant, then Associate, Professor of Science Education, the "and" positionality was less represented amongst both colleagues and the students in many of my classes. This caused me, at times, to wrestle with feelings of a lack of community, identity and a home within science education or education, at large.

I have come to understand my journey in academia as navigating the interstices of multiple disciplines and communities. This process of both abiding by and rejecting traditional boundaries and identities in academia aligns, in part, with Homi Bhabha's (2015) post-colonial theoretical perspective of cultural hybridity and interstitiality. In this chapter, I draw upon Bhabha to reflect on my experiences and outcomes from an interstitial/ "and" positionality in science education and as a Black female academic.

9.1 Hybridity and Interstitiality

Homi Bhabha's (2015) concepts of hybridity and interstitiality are highly relevant to issues of identity, power and place for socially oppressed groups. According to Bhabha, hybridity is a form of power, an "incipient critique" where cultures and identities are democratically negotiated at interstitial points of contact, characterised by both sameness and difference, i.e. "cultural similitude" and "signifiers of alterity" (p. ix). While institutional power structures work to differentially marginalise and constrain groups, hybridity and interstitiality expose those acts of agency and authority from amongst those Other-ed in society, demonstrating how socially oppressed individuals and groups exhibit force in defining their identities in the face of oppression. Everyday instances of discrimination or evidence of systemic inequities are rooted historically and are indicators that contemporary pluralist discourses are insufficiently just. Theoretical lenses of cultural hybridity and interstitiality can bring into focus those "... forms of [critical] agency and affiliation not easily 'named' in the accredited architectures of political power" (Bhabha, 2015, p. xi).

Drawing from Gramsci (1971), Bhabha (2015) argued that evidence of hybridity can be located in the interstices between "organic movements" and "conjunctural phenomena" (p. xii). Organic movements are the ongoing, seemingly natural, societal structures that are, at the same time, contradictory to justice and equity. Bhabha (2015) described them as "incurable" (p. xii), thus implying that they necessitate a radical societal restructuring; however, organic movements are defended robustly by those in present positions of power as status quo societal arrangements have afforded them that power. Conversely, conjunctural phenomena arise from the combination of minor, politically critical forces and target those incurable forms of power. In the interstitial space between organic movements and conjunctural phenomena, the force of cultural hybridity may construct "conjectural phenomena" which relates to "... the contemporary, 'the day-to-day', and to the aspirational future.... '... not the reconstruction of past history but the construction of present and future history....'" (p. xii). Conjectural phenomena are, therefore, agentic challenges to structural oppression and the reauthoring of identities and place within a diverse society; however, the goal is not to attain or reclaim an "authentic" identity representative of the removal of all influence of oppressors, but one "... however mixed, however 'multi', however intersective or intercultural" (p. xii) as a truly equitable and pluralistic society can be.

9.2 Crafting an Interstitial Identity

One such 'multi', intersective representation of my scholarly identity in science education was my *Cultural Studies of Science Education (CSSE)* journal article entitled, "*New Geography for Resistance: The Engagement of Diversity in Doc McStuffins as an Out-of-School STEM Setting*" (Mark, 2022). In this article, I argued that there exists historical racialised and stigmatising ideologies about People of Colour, immigrants and low-income White communities regarding health, education and culture that have been so normalised that they have inadvertently surfaced within the space of a children's television show, which was a form of out-of-school STEM engagement for young children. The insights gained from this article and my conference presentations on the work have been incorporated into my own and other teacher educators' courses, thereby impacting teacher candidates, especially as it relates to implicit biases and dominant ideologies. This article is representative of my interstitial identity, bringing together several disciplinary traditions, theoretical perspectives, and experiences, including teacher education/curriculum and instruction, particularly science/STEM education; counselling psychology and career development across the life span; critical race studies and sociological critiques of health and education; and pop culture, media studies and deconstruction analyses.

Such transdisciplinary scholarship is exciting and eye-opening to me. Even more, the implications are relevant to many diverse fields and applications; however, as stated earlier, I have navigated, and continue to confront, several challenges. I have struggled, at times, to find community, identity and a solid scholarly home with an intersectional/ "and" positionality. I do recognise, though, that these struggles have arisen, in large part, from the traditionally siloed disciplines that structure much of academia, historically and through today. Consequently, the disciplinary interstices lack institutional recognition and individual faculty tend to stay within their own communities, both on their university campuses and within professional research societies; however, that was where I saw great potential for novel and interesting scholarship and teaching.

As a student, I intentionally engaged with several different colleges and departments. When deciding upon electives, I would comb through the course catalogue, covering almost all departments for possibilities. As a faculty member, I similarly planned to be a citizen of the university. For instance, in my early Assistant Professor years, I proposed a new course on Gender and STEM and sought to teach it cross-listed with curriculum and instruction, women and gender studies and biology. Much of the work to populate the course was my responsibility, because these cross-college course offerings were not as normalised as I had expected, I had to rely on word-of-mouth and personal contacts in different colleges and departments to promote the class to their students.

While I did not have the theoretical language at the time, my interstitial identity was intentionally developed; however, with advances in education and in my academic career, my lack of traditional disciplinary fit, especially in science education, made me doubt this interstitial approach. Worse, I began to newly perceive of my "and" positionality as a liability, especially as I navigated the pre-tenure hurdles where faculty are charged with establishing a recognisable scholarly identity; however, Bhabha has argued that all identities are fragmented, hybrid and unstable (Fay & Haydon, 2017). Thus, Bhabha's ideas have guided me back to my original aims and beliefs with greater confidence.

Although I struggled, at times, to find community, identity and home with an intersectional/ "and" positionality, for me, the positives of interstitiality outweighed the negatives. By maintaining my intersectional/ "and" positionality, I have remained intellectually sustained through my research which continues to extend into interdisciplinary spaces. For instance, a new line of my research involves sociological theorising about a youth sports-education complex as a system of racial capitalism and social control, which has built from my previous research on STEM career development amongst Black male students in high-revenue sports, namely basketball and football (Mark & Alexander, 2019). Another new line of research funded by the College of Business at the University of Louisville is focused on positive leadership (Blanch et al., 2016) within a non-profit education and youth development organization working to support autonomy and empowerment amongst immigrant West African students pursuing sport and education in the United States. (Mark & Thomas, 2023). This similarly has built from my earlier research on social justice in and through science/STEM education.

Another positive outcome of my intersectional/ "and" positionality in academia, an important one, has been the ability to form connections with other academics in multiple disciplines and fields. Since, at times, I experienced a lack of recognition and intellectual community within my immediate home of science education because of this intersectional/ "and" positionality, I worked to establish community elsewhere in settings that have been both overlapping and distinct. This, again, has been fulfilling and led to exciting research and teaching outcomes. For instance, I was invited to participate in a National Science Foundation research incubator focused on advancing gender inclusivity in tertiary-level biology education (Eddy et al., 2020). Through this project, I collaborated with almost 20 scholars in biology, women and gender studies, and education throughout North America to critique core textbooks used in introductory undergraduate biology courses, along with other curricula and pedagogical practices, and to devise recommendations for individual, institutional and ideological changes in advance of gender inclusivity and greater critical understanding of sex and gender.

9.3 Challenging Other-Ness

Bhabha (2015) argued that there exists no "pure" culture; no culture untouched or not impacted following interaction. Both cultures are changed by the interaction. Further, by regarding some cultures as more pure than others, an unequal division between the two "Other"-s the latter; what is gained by the former from the latter is discounted and, relatedly, the value of Other-ed identities to exist in and of themselves is denied. Bhabha (1994) further argued the empowering value of performing and contesting identities in "interrogatory, interstitial space[s].... [as this] opens up the possibility of a cultural hybridity that entertains difference without an assumed or imposed hierarchy" (p. 5). Thus, establishing an interstitial home involved, to some extent, resisting *Other*-ness, but also coming to accept and enjoy the lack of an obvious, definitive community. At the same time, maintaining this identity, I was able to produce meaningful interdisciplinary and intersectional scholarship, which, in time, helped me to understand the communities in which I do fit. In line with Bhabha, this home had to be negotiated.

Establishing a home and community ties beyond science education was additionally important because, in times of challenges within one setting, be it my home department or college or a particular research community, my scholarly and social connections could be sustained elsewhere. In times where my value as a faculty member or scholar was being undermined or overlooked, this recognition was found elsewhere. This invaluable advice was imparted upon me by a crucial faculty mentor, a Woman of Colour faculty member and who is a prolific and notable scholar in race, gender, language and education. She has advised me multiple times that it is a mistake for only your home department or institution to recognize you or the value of your work; you must be known elsewhere, especially amongst those in institutions that are more rigorous or competitive than your own. Thus, beyond being sustained intellectually by learning from other fields and disciplines and being challenged to conceptualize problems and solutions in transdisciplinary ways, establishing a home between and across communities has been, at times, protective and affirming.

This faculty mentor's guidance was particularly apt and well-advised as she also embodies an interstitial identity. With a Ph.D. from Harvard University in Anthropology and Education, along with degrees in human development and teacher education, and research in such areas as Black teachers and pedagogy, sociolinguistics and Black women's talk, and urban education and ethnographic analyses of classroom interaction, my faculty mentor and her research is well-known beyond traditional department and disciplinary boundaries. She advised that for education, in particular, it is important to understand the disciplinary structure and culture of departments as some education departments or colleges can be defined very narrowly, for instance, in terms of teacher preparation programs, while some may be defined more broadly to also include education foundations that are more closely related to disciplinary arts and sciences, such as history, anthropology, sociology and psychology. As a scholar with an interstitial identity, the latter better aligns with my interests and expands opportunities for transdisciplinary course offerings and finding scholarly community within my home department.

Protection and affirmation are particularly important for me as a Black female scholar in a predominantly White institution and in the predominantly White field of teacher education (King, 2024). As an Afro-Caribbean immigrant with an audible foreign accent and locs, I stand out amongst my peers and students. Furthermore, I am committed to teaching and research centred on race, power, and social critique. As part of those efforts, I work to unsettle taken-for-granted beliefs and practices regarding race, and how it intersects with multiple other forms of identity, as well as introduce new perspectives. Hybridity and interstitiality affords individuals "... possibilities for local resistance to colonialism" (Fay & Haydon, 2017, p. 29). In education, oppressive racial ideologies and practices are forms of colonizing forces, limiting possibilities for forms of pedagogy and equitable, transformative life outcomes for many Students of Colour and students socially oppressed by their social identities. From an interstitial positionality, drawing from beyond teacher education, I have been able to contribute to challenges raised against such dominant and colonizing forces, disrupting traditionally oppressive schooling and social practices. For instance, given my interests in media studies and deconstruction, I have regularly incorporated or tangentially referenced stand-up comedy bits, novels, hip hop lyrics, documentaries and television procedurals in my classes from Critical Race Theory to Middle and Secondary Science Teaching Methods. Some of these sources, especially some Black stand-up comics, would be typically considered crude, controversial and irrelevant; however, I use these to draw attention to and learn from these public scholars and social critics whose sources of knowledge and lived experiences can, at times, reflect or rival the White intellectual thought and theorising that is traditionally deemed academic.

As an example, in my Critical Race Theory course, I work with students to distinguish liberal perspectives on race as opposed to our targeted critical perspectives on race. To this end, I draw on excerpts from Chris Rock's Netflix Special, *Tamborine*, in which Rock incisively and tangibly critically analyses issues of race and power in the United States. historically and contemporary. In identifying the permanence of racism in U.S. society, a core Critical Race Theory tenet, Rock explained that, in today's society, we might not see the *Jim Crow* era signs of barring Black people from entering certain buildings and facilities; but the contemporary economic system works just as efficiently. Rock stated:

> But racism is not gonna stop. It ain't never stopping. It's the American way, man. You know, they used to have signs up that said, "Whites only." They used to have signs up that said, "No Blacks allowed." Now they got something new. It's called prices. That's right. Prices are the new Jim Crow. You know, the Four Seasons hotel does not say, "No Blacks allowed." But a $4000-a-night hotel suite sure does. That ought to handle your n**** problem right there. Prices are the new Jim Crow, baby. You know, Whole Foods does not say, "No Blacks

allowed," but a $7 orange sure does. It says that they don't want my Black a** in here. A couple of White people are going, "They don't want us in here either."

Derrick Bell, a founding scholar of Critical Race Theory, argued that "racial equality is, in fact, not a realistic goal" (Bell, 1992, p. 363); that the United States was founded on principles of racism and these are structured into the functioning of society; that they remain present, even if their forms change:

> Today, one can travel for thousands of miles across this country and never come across a public facility designated for "Colored" or "White." Indeed, the very absence of visible signs of discrimination creates an atmosphere of racial neutrality that encourages whites to believe that racism is a thing of the past. Today, blacks experiencing rejection for a job, a home, a promotion, anguish over whether race or individual failing prompted their exclusion. Either conclusion breeds frustration and eventually despair. We call ourselves African Americans, but despite centuries of struggle, none of us – no matter our prestige or position – is more than a few steps away from a racially motivated exclusion, restriction or affront. (Bell, 1992, p. 374)

To preserve our morale and sanity in the face of ongoing racism and to resist in more effective ways, Bell explained that we should adopt a perspective of racial realism. One illustration of this racial realism was Mrs. MacDonald who took risks and engaged defiantly with the goal to "harass white folks" (Bell, 1992, p. 378) who "... she well knew, held all the economic and political power, and the guns as well.... [but] she determined to resist her oppression, [and] she was triumphant" (pg. 378). Rock skilfully and tangibly argued the tenet of racial realism and his artistic engagement was an embodiment of that harassing defiance. As a Black woman faculty in predominantly White academia in the United States, I understand that transformative institutional change is slow and hard to achieve. In the face of that racial realism, creating spaces of joy and celebrating resistance by integrating lesser-heard narratives and diverse forms of expression has been one avenue I have traversed to persevere.

Through vivid storytelling and memorable metaphors, Rock and many other artistes demonstrate the intellectual prowess of academics and traditional (White) scholars. For example, in *Kill the Messenger*, Rock said, "The Black man gotta fly to get something the White man gets to walk to" as a metaphoric criticism of abstract liberalism as a dominant ideology that denies structural privileges and barriers to autonomy and self-determination in society based on racial differences.

Not only have I worked to decolonise (science) education ideologies and practices, but I have had to work against becoming colonised myself. Again, as a Black female academic in teacher education, I am easily physically and culturally distinguished from the vast majority of both my peers and the students in my classes. Due to this positionality, I have experienced unique challenges, especially in the area of teaching proficiency, while on the tenure track. The populations of both teacher educators and teacher candidates are predominantly middle- to upper-middle class White females (King, 2024). Many classroom teachers advance into teacher educator roles in preparation programs. Thus, many faculty members in teacher education programs are culturally similar to students in their classes. As such, cultural interactional patterns and life experiences are more likely to be similar between the two groups, thus aiding in establishing positive rapport and relationships that may translate into positive teaching evaluations. Additionally, many teacher candidates lack diverse cultural experiences, especially meaningful interracial interactions, including very few, if any, faculty of colour in their courses.

In contrast to the racial, gender and cultural similarities between most of my peers and students in teacher education courses, I am a Black female professor and research evidence has shown that Black female professors tend to get relatively low student evaluations compared to their peers (Chávez & Mitchell, 2020). Furthermore, immigrant professors with particular audible accents can also get poorly evaluated relative to their peers (Lawless & Chen, 2017). Beyond my racial and cultural identities, most of my courses are explicitly focused on potentially controversial and political topics, such as race, identity, power and privilege, which can elicit negative student reactions or poor evaluations of teaching. Teaching such material requires careful planning and execution (Mark, 2021b). Successfully engaging predominantly White students in critical race studies that challenge Whiteness involves carefully managing emotions such that students do not defensively revert to Whiteness ideology where they project anger and fear onto People of Colour rather than work through their genuine feelings in response to learning about White supremacy and social injustice (Matias & Mackey, 2016). These factors have created unique challenges to me in my department that I had to overcome with grace in order to not present as blaming others, yet prove to be an effective teacher. As an example, a former department chair inquired into the nature of a book study in one of my courses since another White female colleague taught the same book in other sections of the course, but the student feedback involving me was racially charged.

There are no institutionally established means by which to confidently account for these underlying racial and cultural dynamics within the design of teaching evaluations themselves. You must provide the evaluations to tenure and promotion review committees as is, thus committee members can interpret these independently. You can attempt to provide some context or explanation for any relatively low evaluations in your personal narrative that accompanies your teaching evaluations and other evidence; however, you run the risk of appearing to complain or blame others, thus creating a negative impression of yourself to the committee. You can work to offset the teaching evaluations by including other evaluations by peers; however, while peer evaluations might be well-meaning, they may still potentially be filtered through experiences and distinct cultural interactional patterns of peers who are different from you. Therefore, any problematic student teaching evaluations are not improved, but rather couched within other forms of evidence. Furthermore, the work to gather such secondary forms of evidence is an additional burden endured by female scholars of Colour.

So, in working towards higher teaching evaluations, in the first place, or seeking strong secondary evaluations from peers, I have experienced, as a colonizing force by the institution of teacher education, in particular, and academia, more broadly, implicit and explicit expectations to adapt the teaching and interactional practices of predominantly White female faculty peers in order to garner comparable teaching evaluations to peers in my department, while not being sufficiently supported in accounting for the racialised context of my evaluations. While adopting an interstitial positionality does not improve one's teaching evaluations or make the process of earning higher teaching evaluations easier, it does help establish a mindset capable of enduring the inequity of racialised teaching and tenure evaluations in academia. As Bhabha has argued, the interstitial space is one in which oppressive, colonising forces can be challenged. From a critical realist perspective (Bell, 1992), structural racism exists in this academic tradition of pursuing tenure and promotion; however, rather than cave to domesticating politics (Rodriguez, 2005), "... the hybrid strategy or discourse opens up a space of negotiation where power is unequal but its articulation may be equivocal.... neither assimilation nor collaboration" (Bhabha, 2013, p. 58). This position demands a negotiation in the interstices amongst cultures such that both socially oppressed and privileged identities are valid and recognized without erasing the influence of any culture/s on others. Rather, than aim to be subsumed by science education, teacher education or academia, broadly, these struggles are indicative of a resistance to the "mummification" of culture (Tate, 2000).

9.4 Interstitiality in Science Education Scholarship

From an interstitial perspective, I have been intrigued by and pursued a number of wide-ranging questions that have centred on theorizing about race, ethnicity, gender, power and oppression in educational settings. Sociology and anthropology have been highly informative disciplines in shaping my epistemological and paradigmatic approaches to scholarship. Earlier on, during my doctoral studies, I was mentored by two faculty members, Professors of Science Education and Counselling Psychology, respectively. This exposed me to the learning sciences and counselling psychology as it relates to science/STEM learning and career development, particularly in urban educational settings. For instance, the bulk of my research experiences focused on high school-level weekend and school vacation STEM camps in which students engaged in immersive, hands-on learning activities, e.g. (Blustein et al., 2013; DeBay et al., 2012; Mark, 2016; Mark et al., 2013). These STEM activities additionally involved features intended to grow students' interest in, knowledge of and plans to pursue STEM-related careers. Such features included roundtable dialogues with diverse STEM career professionals, conducting research on specific STEM careers, and completing self-assessments of their career interests and personality traits.

Most meaningful to me was that, at the core of these learning and career development interventions, were critical perspectives on education and working. The approach to growing young people's interests in and pursuit of STEM education and careers was not in advance of competitive nationalist economic interests (Basile & Lopez, 2015), but rather as a means for young people, primarily from under-served racial, ethnic and economic backgrounds to overcome historical barriers to high quality STEM education and to be able to compete for STEM careers, if they choose to pursue these career pathways. Professional STEM careers were perceived as a resource to economic and social advancement for young people who were not already indicating much interest in, awareness of or educational and career plans in STEM. Rather than help the already most prepared and competitive students in STEM, e.g. those in advanced and honours STEM courses, acquire additional resources, these weekend and school vacation STEM camps were intended to tailor to the targeted student audience's interests and priorities, which were social justice, equity and peer support.

Similarly, I was ushered into a critical approach to career development and working. In one of my first meetings with my faculty mentors, the professor of counselling psychology handed me a copy of an article he had published with some of his collaborators, focused on an emancipatory and community-centred approach career development theory, research, and practice. The centrality of an emancipatory framework and the integration of theory and practice, in particular, were compelling to me and aligned with my priorities. Later, in auditing the counselling psychology department's career development doctoral course, I read Blustein's (2006) *The Psychology of Working: A New Perspective for Career Development, Counseling, and Public Policy*, which became a central theoretical framework for my dissertation and later a journal article (Mark, 2016). As with the critical and inclusive orientation of the STEM education and career development camps, a Psychology of Working perspective centred historically overlooked and excluded populations from counselling psychology research and clinical practice, namely populations from working-class backgrounds, People of Colour and non-college-educated populations. Instead, this framework was sensitive to the ways in which structural factors can constrain people's opportunities to acquire career experiences that are self-determined and empowering. This framework attended to the oppressive realities of many populations in pursuing work for survival and identity development.

Psychology has been critiqued as a discipline for focusing too heavily on individual processes of cognition and meaning-making and less so on the structural organization of society (Bryant-Davis, 2021; Carter, 2007; Torres Rivera, 2020). Thus, a Psychology of Working framework permitted use of a lesser represented critical perspective in the discipline, which was what I was seeking. Mark (2016) was limited, however, by relying on interview data. Missing from this research was the complexity of human interaction in negotiating meaning in cultural contexts, which required deeper immersion in the community setting. For this, I drew upon anthropology for communities of practice theory (Lave & Wenger, 1991; Wenger, 1999). Again, earlier in my doctoral studies, my doctoral adviser recognized my inclinations to anthropological theory and guided me towards Holland et al.'s (2003) *Identity and Agency in Cultural Worlds*. These anthropological sources explicated social processes in which people negotiated meaning and identity through practice, but importantly for me, these theories acknowledged diversity and power differentials in discursive processes.

However, having entered educational studies for the purpose of inquiring into how racial, ethnic, national, gender and other socially constructed differences shaped U.S. education, work and other aspects of society, what I still needed were robust theories on race, racism and intersectional forces at a macrolevel that could then explain group and individual level behaviour. This, I gained in a Critical Race Theory course in which I enrolled at Tufts University's School of Arts and Sciences. Although a doctoral student at Boston College, I was able to enrol in courses at Tufts and other neighbouring universities through a graduate consortium. From this course and career-long study, I developed an understanding of race, racism and intersectional oppressions as institutionally structured and ideologically maintained in order to preserve exclusive privileges to individuals racialized as White (Crenshaw et al., 1995; Delgado & Stefancic, 2017).

Critical race theory was the enduring theoretical lens for me moving forward both in academia and in life, in general. I now teach a course on critical race theory and use it in my scholarship, e.g. (Foster et al., 2024; Mark, 2021a, 2022). Critical race theory is excellently suited as an interstitial theoretical framework. Originating in the field of law, critical race theory, along with its offshoots, is actively engaged in scholarship, teaching and praxis in a wide range of disciplinary domains, such as English literature (Engles, 2006), sport management (Cooper et al., 2017), and geospatial analysis (Vélez & Solórzano, 2017). This last application is particularly interesting to me given my past experiences in supporting young people develop skills in the use of geographic information systems technology for community land development projects as a means of critical STEM education and youth development (DeBay et al., 2012). In its home field of law, critical race theory has expanded beyond criminal justice and has been advocated in the sub-domain of taxation (Shurtz, 1997). Despite seeming too remote for a science education scholar, each of these have been incorporated into my research and teaching. Currently working at the interstices of the sociology of race and ethnicity, education, sport and science culture and practices, I continue to use critical race theory centrally in my scholarship. To this, I have added crucial sociological theoretical concepts such as racial capitalism and ideological social control, among others. I will close by summarizing two such forthcoming projects next. My motivation in pursuing these and other interrelated projects originate from my own and my family's educational and sports journeys in the racialised labour context of U.S. intercollegiate sports. This sport and education complex, which I expand upon below, spans the globe, implicating systems of immigration, as well. It is interesting and insightful to reflect upon the changing narratives and policies regarding paying athletes and/or allowing athletes to profit financially from their involvement in NCAA sports, especially having been an NCAA athlete during the period in which neither of these options were entertained.

9.4.1 Gendered Racial Capitalism in a Youth Sports-Education Complex

In an earlier publication, I argued that, for high-revenue Black male athletes, i.e. high school and college, but increasingly middle and elementary, basketball and football athletes, their academic/intellectual development is, at best, overlooked, and, at worst, wilfully sacrificed in order to expropriate their physicality for capital gain in markets that span sports at the youth, college, and professional levels and K-16 education (Mark & Alexander, 2019). In this project, I focused on Black male athletes as a population underrepresented in STEM college majors and robust STEM education as preparation for high quality STEM careers. These are problems of educational neglect, along with career under-development in STEM and occupational segregation for Black males. I have further developed this argument to be that this system is a form of modern-day ideological social control (Jackman, 1994) in a gendered racial capitalist complex of youth sports and education. This argument centres on a theoretical model of a youth sports-education complex as a system of racial capitalism (Cox, 1948) that exploits all athletes, but disproportionately harms Black athletes. The oppression and expropriation is, furthermore, intersectionally gendered (Crenshaw, 1991) as these forces are particularly concentrated on Black male athletes who are most representative of the revenue-generating sectors of the complex (Canada et al., 2022; Crawford, 2019). Using this model of gendered racial capitalism, I then aim to characterise the specific ways that Black male athletes have been commodified and controlled within a youth sports-education complex as a system of gendered racial capitalism. This research has implications for science educators, along with educators across disciplines and the various professionals who work with students and athletes.

In this project, I define the youth sports-education complex as spanning K-16 education, which is defined as the period beginning with kindergarten, typically started at five years old in the United States, to the traditional four years of undergraduate college education. It should be noted that in the case of college athletes, especially regarding Black athletes in high-revenue sports, many take longer than four years to complete their athletic tenures, which is four years (or seasons) of active competition in the NCAA, which can include breaks due to injury, and take longer than the traditional four years to complete their degree requirements. Additionally, although athletes in college are 18 years and older, the term "youth sports" is used for two reasons. First, "youth" is meant to signify that the athletes are minors from the start and for the majority of their involvement in the sports-education complex of focus in this article and are students enrolled in education institutions during this involvement. Second, the word "amateur" or "student-athlete" is not used to distinguish the aforementioned stage of sports participation from the "professional" income-earning stage of sports that typically follows the end of NCAA involvement. The word "amateur" or "student-athlete" is not used since these are narratives of racial and social control manufactured and utilized by the NCAA over its institution athletes and have historically been used to prevent athletes from earning income from their labour as NCAA athletes until recent changes in 2021 (Burroughs et al., 2023; Canada et al., 2022; Staurowsky & Sack, 2005).

9.4.2 *Racialised Student Labour in Intercollegiate/NCAA Sports*

Intercollegiate sports were initially established as student-led participatory games and activities (Sack & Staurowsky, 1998). These later developed into local and live spectator events. In turn, intercollegiate sports have since progressed into a major corporate entertainment enterprise or show business with a largely national, television-based viewing audience and driven almost entirely by football and men's basketball (Duderstadt, 2009). These sports are disproportionately representative of Black athletes from low-income backgrounds (Edwards, 1984; Hawkins, 2010; Runstedtler, 2018; Smith, 2014). Some critics of this commercialisation in intercollegiate sports blame forces external to colleges and universities, such as the aforementioned beverage companies and television/broadcasting companies; however, high-profile university coaches are major factors who have distorted the field of intercollegiate sports by navigating it as a means to become wealthy (Sperber, 2000).

Paradoxically, this billion-dollar intercollegiate athletics enterprise is comprised of not-for-profit higher educational institutions and governed by the National Collegiate Athletics Association (NCAA), a not-for-profit organization. In the face of this hypocrisy, the NCAA has held firm to their narrative construction of the "amateur student-athlete," which served to restore the image of the association and its intercollegiate competitions in the wake of rampant corruption, gambling and excessive competitiveness (Duderstadt, 2009; Staurowsky & Sack, 2005). A major function of this "student-athlete" narrative was to characterise the association's intercollegiate competitions and the athletes involved as "amateurs" as opposed to professional activities and labourers, respectively. This would work to legitimise their paternalistic control over athletes, for instance restricting their earnings and their decision-making, e.g. ability to transfer to different institutions, yet withholding critical protections normally afforded employees, such as workman's compensation for serious or career-ending injuries (Burroughs et al., 2023; Sack & Staurowsky, 1998). Additionally, any efforts by intercollegiate athletes or advocates towards unionising have been delegitimised on the basis that the athletes and the NCAA were engaged in "amateur" gameplay. Instead of wages, athletes received a minimum compensation for their labour in the form of athletic scholarships; however, although athletic scholarships were once four-year guarantees to college athletes such that they can finance their college enrollment, under pressure from coaches, in 1973, the NCAA changed its athletic scholarship policy to one-year renewable grants-in-aid (Duderstadt, 2009). Coaches felt restricted by athletes being guaranteed four years of financial support as they were unable to dismiss or withhold funding from those athletes who did not contribute to team wins and their budget to recruit new prospects was similarly restricted until veteran athletes graduated, withdrew or otherwise left the team or university (Sperber, 2000).Thus, this was a major change intended to grant coaches much greater control over their athletes based on their athletic labour value.

Du Bois (1964) located Black labour as foundational to the development of the United States as a nation, so much so that Black labour in the form of slavery was the unacknowledged cause of the Civil War. Following the Civil War, the surplus-value of Black labour in the United States remained a necessary resource for capital accumulation, which had implications for land, labour and educational policies regarding Black Americans. For people in the African Diaspora, colonialism and slavery have been significant institutions in shaping ongoing life experiences and chances. These patterns have continued throughout history. Black college athletes in commercialised intercollegiate sports are one of several contemporary sources of labour for capital accumulation and, as such, policy development have been similarly influenced by race intersecting with economic interests, i.e. racial capitalism (Matlon, 2016). This historical and political insight presents significant tensions in how to resolve my own and peers' exploitation, especially Black male family and friends situated in the most commericalised core of NCAA and NBA men's basketball. Even more, in parenting, community engagement and education, I struggle to resolve the cognitive dissonance of the dueling forces between: (i) bearing witness to my children's and other youth's genuine joy, talent and development in various sports, along with the opportunities sports participation can provide in accessing higher education; and (ii) the legacy of enslavement and expropriation of Black labour, coupled with the strategic underdevelopment within Black communities as means of social control and oppression. Furthermore, as a mother of daughters who enjoy and participate in sports, I recall Sack and Staurowsky (1998) lamenting how the integrity of women's athletics declined as they similarly became increasingly commercialised as men's sports.

Du Bois' sociological undertakings, such as in *Black Reconstruction, The Souls of Black Folk* and *The World and Africa*, were, furthermore, ways of challenging colonial forms of knowledge (Itzigsohn & Brown, 2020), i.e. systems of belief and schemas for interpreting the world driven by those in positions of power and used for legitimising and reinforcing structural arrangements in society. Applying Du Bois' critical and reconstructionist analytical lens, through this study, I aim to challenge dominant, controlling conceptions of Black athletes and newly interpret their roles and experiences within intercollegiate sports as a system of racial capitalism. The implications for this study include education theory, policy studies and teacher education as it relates to the ways in which students and the education they are afforded are racialised. It aims to elucidate the structural forces shaping the educational experiences and life chances of Black youth, locating these forces in their historical roots and tracking their contemporary development. It is important for pre-service and in-service teachers, along with others involved in the education of Black youth, to understand this sociopolitical context surrounding their Black students, especially Black males, in order to grasp the ramifications of their engagements with these youth as educators, mentors and societal gatekeepers. I was initially drawn to the scholarly study of science education as I recognized science as a culture stratified with power dynamics. I was driven to elucidate the long-established and evolving ways in which the privilege to access science as a student, as well as the quality of one's science education, has been tied to structural racism and other social formations. The underrepresentation of Students of Colour, especially Black males, in advanced science education and professional careers is situated in this larger sociopolitical context.

9.5 Conclusion

Both intentionally and inadvertently, I have crafted an intersectional and interstitial home for myself in academia. Interstitiality served as a framework through which I have worked and connected across disciplines and fields such as critical race studies, sociology, teacher education and science/STEM. An interstitial home was personally and professionally important to me, satisfying my transdisciplinary interests and inclinations. Interstitiality also helped me connect with multiple overlapping and distinct communities to advance my teaching and scholarship, as well as to attain affirmation and protection. These benefits endured although the interstices often lack institutional recognition and clearly defined communities. Very important, as well, establishing an interstitial home aided me, to some extent, in resisting Other-ness. By theoretically understanding my positionality through the lens of interstitiality, I have come to understand and situate my efforts in a larger field of work committed to decolonising dominant ideologies in education and society-wide. Finally, as a Black female academic, interstitiality guides me in resisting institutional pressures that would alter the critical focus and methods of my scholarship and teaching; thus, resisting becoming colonised myself.

References

Basile, V., & Lopez, E. (2015). And still I see no changes: Enduring views of students of color in science and mathematics education policy reports. *Science Education, 99*(3), 519–548. https://doi.org/10.1002/sce.21156

Bell, D. (1992). Racial realism. *Connecticut Law Review, 24*(2), 363–379.

Bhabha, H. (2015). Foreword. In P. Werbner & T. Modood (Eds.), *Debating cultural hybridity: Multicultural identities and the politics of anti-racism* (2nd. critique influence change ed.). Zed Books.

Bhabha, H. K. (1994). *The location of culture*. Taylor & Francis. https://books.google.com/books?id=WjbbCgAAQBAJ

Bhabha, H. K. (2013). Culture's in between. In S. Hall & P. D. Gay (Eds.), *Questions of cultural identity* (pp. 53–60). SAGE Publications.

Blanch, J., Gil, F., Antino, M., & Rodríguez-Muñoz, A. (2016). Positive leadership models: Theoretical framework and research. *Psychologist Papers, 37*(3), 170–176. https://www.redalyc.org/articulo.oa?id=77847916003 (IN FILE)

Blustein, D. L. (2006). *The psychology of working: A new perspective for career development, counseling, and public policy*. Lawrence Erlbaum Associates.

Blustein, D. L., Barnett, M., Mark, S., Depot, M., Lovering, M., Lee, Y., Hu, Q., Kim, J., Backus, F., Dillon-Lieberman, K., & DeBay, D. (2013). Examining urban students' constructions of a STEM/career development intervention over time. *Journal of Career Development, 40*(1), 40–67. https://doi.org/10.1177/0894845312441680

Bryant-Davis, T. (2021). *Liberation psychology: An overview* (Exploring Liberation Psychology: Teaching, Training, and Practice, Issue. https://www.youtube.com/watch?v=I16-jzarVns

Burroughs, B., Cummings, K., Johnson, R., & Romney, M. (2023). The Rinaldi frame: The NCAA, College Gameday, and the commodification of Black hardship. *Sport in Society*, 1–16. https://doi.org/10.1080/17430437.2023.2235302

Canada, T., Pericak, K., & Seward, M. D. (2022). Amateurism as a narrative of control: An interdisciplinary approach to the lived experiences of college athlete. *Sports Innovation Journal, 3*(SI), 55–68. https://doi.org/10.18060/26003

Carter, R. T. (2007). Racism and psychological and emotional injury: Recognizing and assessing race-based traumatic stress. *The Counseling Psychologist, 35*(1), 13–105. https://doi.org/10.1177/0011000006292033

Chávez, K., & Mitchell, K. M. W. (2020). Exploring bias in student evaluations: Gender, race, and ethnicity. *PS: Political Science & Politics, 53*(2), 270–274. https://doi.org/10.1017/S1049096519001744

Cooper, J. N., Nwadike, A., & Macaulay, C. (2017). A critical race theory analysis of big-time college sports: Implications for culturally responsive and race-conscious sport leadership. *Journal of Issues in Intercollegiate Athletics*(10), 204–233. http://csri-jiia.org/wp-content/uploads/2017/12/RA_2017_11.pdf

Cox, O. C. (1948). *Caste, Class, and Race: A Study in Social Dynamics.* Modern Reader Paperback. https://books.google.com/books?id=AGXaAAAAMAAJ

Crawford, C. (2019). Athleticising young black lives: confusing education with commerce in high school sports. In S. Dagkas, L. Azzarito, & K. Hylton (Eds.), *'Race', youth sport, physical activity and health: Global perspectives* (1st ed.). Routledge.

Crenshaw, K. (1991). Mapping the margins: Intersectionality, identity politics, and violence against women of color. *Stanford Law Review, 43*(6), 1241–1299. https://doi.org/10.2307/1229039

Crenshaw, K., Gotanda, N., Peller, G., & Thomas, K. (Eds.). (1995). *Critical race theory: The key writings that formed the movement.* New Press.

DeBay, D., Haley, J., Mark, S., Barnett, M., Anderson, A., Strauss, E., Cotter-Hayes, L., Blustein, D., & Wong, C. (2012). Engaging youth in visualizing sustainable urban plans using geographic information systems coupled with computer visualization. In A. Wals, P. Corcoran, & H. Brandon (Eds.), *Learning for sustainability in times of accelerating change.* Springer.

Delgado, R., & Stefancic, J. (2017). *Critical race theory: An introduction* (3rd. ed.). New York University Press.

Du Bois, W. E. B. (1964). *Black reconstruction in America: An essay toward a history of the part which black folk played in the attempt to reconstruct democracy in America, 1860–1880.* Meridian Books.

Duderstadt, J. J. (2009). *Intercollegiate athletics and the American university: A university president's perspective.* University of Michigan Press.

Eddy, S., Fuselier, L., Jarosi, S., & Casper, A. (2020). *Creating a more inclusive undergraduate biology curriculum* [Grant]. USA.

Edwards, H. (1984). The Black 'dumb jock': An American sports tragedy. *The College Board Review, 131*, 8–13.

Engles, T. (2006). Connecting white noise to critical whiteness studies. In T. Engles & J. N. Duvall (Eds.), *Approaches to teaching DeLillo's white noise* (pp. 63–72). Modern Language Association.

Fay, S., & Haydon, L. (2017). *An analysis of Homi K. Bhabha's the location of culture.* Routledge. https://books.google.com/books?id=2EkrDwAAQBAJ

Foster, M., Mark, S. L., & Baize, J. (2024). Introduction special issue: Educational equity: Cultural and ethnic diversity in schools. *Education Sciences, 14*(1), 38. https://doi.org/10.3390/educsci14010038

Gramsci, A. (Ed.). (1971). *Selections from the prison notebooks.* Lawrence & Wishart.

Hawkins, B. (2010). *The new plantation.* Springer.

Holland, D., Lachicotte Jr., W., Skinner, D., & Cain, C. (2003). *Identity and agency in cultural worlds.* {Harvard University Press}. citeulike-article-id:691663

http://www.amazon.ca/exec/obidos/redirect?tag=citeulike09-20&path=ASIN/0674005627

Itzigsohn, J., & Brown, K. L. (2020). Racial and colonial capitalism. In *The sociology of WEB Du Bois* (pp. 62–96). New York University Press.

Jackman, M. R. (1994). *The velvet glove: Paternalism and conflict in gender, class, and race relations*. University of California Press. http://ark.cdlib.org/ark:/13030/ft958009k3/

King, J. E. (2024). AACTE Update: Teacher preparation program trends 2010–11 to 2020–21. https://aacte.org/resources/research-reports-and-briefs/teacher-preparation-program-trends-2010-11-to-2020-21/

Lave, J., & Wenger, E. (1991). *Situated learning: Legitimate peripheral participation*. Cambridge University Press.

Lawless, B., & Chen, Y.-W. (2017). Multicultural neoliberalism and academic labor: Experiences of female immigrant faculty in the U.S. academy. *Cultural Studies ↔ Critical Methodologies, 17*(3), 236–243. https://doi.org/10.1177/1532708616672688

Mark, S. L. (2016). Psychology of working narratives of STEM career exploration for non-dominant youth [journal article]. *Journal of Science Education and Technology, 25*(6), 976–993. https://doi.org/10.1007/s10956-016-9646-0

Mark, S. L. (2021a). High-revenue athletics institutional policies and practices aligned with key resources for STEM career development. *Journal of Higher Education Athletics and Innovation, 8*(2020). https://doi.org/10.15763/issn.2376-5267.2020.1.8.113-139

Mark, S. L. (2021b). Preparing for inclusivity and diverse perspectives on social, political, and equity issues in higher education. *College Teaching, 69*(2), 78–81. https://doi.org/10.1080/87567555.2020.1820433

Mark, S. L. (2022). New geography for resistance: The engagement of diversity in Doc McStuffins as an out-of-school STEM setting. *Cultural Studies of Science Education, 17*(3), 955–981. https://doi.org/10.1007/s11422-021-10096-y

Mark, S. L., & Alexander, O. (2019). A critical race case study exploration of two Black male student-athletes' STEM career development. *Journal for the Study of Sports and Athletes in Education, 13*(2), 116–145. https://doi.org/10.1080/19357397.2019.1633508

Mark, S. L., DeBay, D., Zhang, L., Haley, J., Patchen, A., Wong, C., & Barnett, M. (2013). Coupling social justice and out-of-school time learning to provide opportunities to motivate, engage and interest under-represented populations in STEM fields. *Career Planning and Adult Development Journal, 29*(2), 93–105.

Mark, S. L., & Thomas, C. N. (2023). Positively leading immigrant west african athletes and families in the global sport arena. *2023 Fellowship Research Report*. https://positiveleadership.louisville.edu/wp-content/uploads/2024/01/Rechter-Fellowship-Final-Report_Sheron-Mark-2023.pdf

Mark, S. L., Trzaskus, M., Archer, L., & Azmani, P. (2022). Fostering social connectedness and interest in science through sports. In A. J. Rodriguez & R. Suriel (Eds.), *Enacting cross-cultural science/stem education research against the odds: A letter in eight chapters to funding agencies, research journal editors, reviewers and policy makers*. Springer.

Matias, C. E., & Mackey, J. (2016). Breakin' down whiteness in antiracist teaching: Introducing critical whiteness pedagogy. *The Urban Review, 48*(1), 32–50. https://doi.org/10.1007/s11256-015-0344-7

Matlon, J. (2016). Racial capitalism and the crisis of black masculinity. *American Sociological Review, 81*(5), 1014–1038. https://doi.org/10.1177/0003122416658294

Rodriguez, A. J. (2005). Rejecting mediocrity and the politics of domestication. *Journal of Curriculum Theorizing, 21*(3), 47–59.

Runstedtler, T. (2018). More than just play: Unmasking black child labor in the athletic industrial complex. *Journal of Sport and Social Issues, 42*(3), 152–169. https://doi.org/10.1177/0193723518758458

Sack, A. L., & Staurowsky, E. J. (1998). *College athletes for hire: The evolution and legacy of the NCAA's amateur myth*. Bloomsbury Publishing USA.

Shurtz, N. E. (1997). Critical tax theory: Still not taken seriously. *NCL Rev., 76*, 1837.

Smith, E. (2014). *Race, sport and the American dream* (3rd ed.). Carolina Academic Press.

Sperber, M. (2000). *Beer and circus: How big-time college sports has crippled undergraduate education*. Macmillan.

Staurowsky, E. J., & Sack, A. L. (2005). Reconsidering the use of the term student-athlete in academic research. *Journal of Sport Management, 19*(2), 103–116.

Tate, A. (2000). Homi Bhabha: The process of creating culture from the interstitial, hybrid perspective. https://rowenasworld.org/essays/newphil/bhabha.htm

Torres Rivera, E. (2020). Concepts of liberation psychology. In L. Comas-Díaz & E. T. Rivera (Eds.), *Liberation psychology: Theory, method, practice, and social justice.* (pp. 41–51). American Psychological Association. https://doi.org/10.1037/0000198-003

Vélez, V. N., & Solórzano, D. G. (2017). Critical race spatial analysis: Conceptualizing GIS as a tool for critical race research in education. In *Critical race spatial analysis* (pp. 8–31). Routledge.

Wenger, E. (1999). *Communities of practice: Learning, meaning, and identity.* Cambridge University Press.

Chapter 10
The 3 C's to My Success in Science Education: Connection, Communication, and Community

Angelia Reid-Griffin

10.1 Introduction

Science education is an area of teacher education that focuses on teaching learners about science and STEM concepts. As with all areas of teacher education, one must have a true commitment to helping others to learn. This chapter will share my journey to become a science teacher and the events that helped to shape my vision and success as a science educator. Growing up in a small rural community in North Carolina, I have always had a love for science, but teaching was not a part of my career plans. Although as a young child, I did spend time playing school with my cousins and friends, teaching our stuffed animals and dolls lessons. But my favorite times of my childhood were spent exploring the riverbank and wooded areas in my aunt's backyard with my cousins. I loved exploring the area with my cousins as our family fished and gathered crabs for our family gatherings. My cousins and I would play hours on end, with the slider turtles and conduct "kid" dissections on any of the dead animals that we found lying along the riverbed. ("Kid" dissections is the term I used to describe how inexperienced children examine dead animals. Using only sticks and plastic utensils to examine the specimen and create their own interpretation of what caused the animal to die). Engaging in play science activities such as this influenced my career plans to be a scientist or medical doctor when I entered college. Once I entered college, I encountered some unique challenges that allowed me to question whether these were my only career options in science. While working as a substitute teacher during my academic journey I was able to see how I could share my love for science with others. This changed my career path to science education.

In many of my science content and science education courses, I was one of the few students of color. As an African American female and the only one in many

A. Reid-Griffin (✉)
UNC Wilmington, Wilmington, USA
e-mail: reidgriffina@uncw.edu

M. M. Atwater and B. A. Garvin (eds.), *Creating and Maintaining a Home in Science Education Research*, Contemporary Trends and Issues in Science Education 65,
https://doi.org/10.1007/978-3-031-99382-4_10

of these courses, I had to be resilient in pursuing my career goals of science and science education. Noted scholars Morris (2016), Howard-Vital (1989), and Bartman (2015) reveal in their studies that African American women pursuing careers where they are underrepresented are characterized as lacking the academic training of their counterparts (Haynes, 2019). These are indeed some of the unique challenges I faced as I completed my degrees in science and science education. Having a strong family foundation was essential in helping me to feel confident about myself and to stay focused on my career goals. By acknowledging the racial and gender challenges and working toward solutions to accomplish my goals, then I was able me to move closer to my space where I am valued and always included. I hope this body of work will serve as a testament to the importance of perseverance and the impact of having a supportive network.

Education holds a place of paramount importance, serving as both a cherished value and a cause for celebration in my family. This prioritization of learning transcends mere academic achievement, embedding itself in the very fabric of familial identity and legacy. Many of my family members worked in this field, including my mother. This had a significant influence on my decision to study science and later pursue degrees in science education. I often consider it to be predestined for me to enter a career in teaching with many of my relatives employed in education. Despite my earlier career plans to not become a teacher, I found myself wanting to teach. In my science courses where I was one of a few African American students, I often wondered if I even belonged. I knew I belonged and that I needed to be a teacher to help other children like me know that they belong too. Education was not just an individual pursuit for me but a communal endeavor where I was able to share my successes and challenges with my family.

My education has been in mostly predominantly White institutions (PWIs). In these institutions, I struggled with adjusting to college due to the lack of engagement and sense of belonging as one of a few African American females in my degree programs. I credit the values of my family and my supportive networks in providing me the confidence to "navigate stereotypes and dynamics" (Shavers & Moore, 2014) about my ability to be a scientist and science educator. My parents and other relatives instilled in me that I should never let anyone limit my abilities. They always said, "Do your best and do not let anyone tell you are any less." I continue to follow this belief today as I continue to work in spaces where I am one of a few African American women in science education. In Haynes's (2019) research she addresses how African American women like myself, have faced unique challenges while attending predominately White institutions. Learning how to navigate in these "paradoxical spaces" as a strategy to be successful in these spaces where I was often ignored or devalued (Haynes, 2019). Experiencing first-hand judgments on your abilities and identity, by peers as well as instructor, shows how these academic spaces test your resilience. Attending White campuses as African American women can offer a unique set of challenges and adjustment issues. The lack of representation in these institutions results in feelings of isolation or being misunderstood. This dissonance extends beyond the classroom, affecting participation in program activities, seeking mentorship, and harming mental health. The practice of having to constantly prove one's worth is exhausting and mentally draining. In addition to having to adapt to fit into these predominantly White spaces is disheartening.

The success of African American females on predominantly White campuses underscores the importance of diversity and inclusion efforts within higher education. Institutional changes are needed across the globe to advocate for policies that address educational inequities, provide resources such as scholarships and financial aid tailored to assist students from historically marginalized communities, and create platforms outside the classroom and campus environment to empower their voices. Addressing these feelings of marginalization requires a concerted effort at all levels of the academic structure. Institutions must commit to fostering diversity, equity, and inclusion, not just in their mission statements but in their everyday practices. This includes acknowledging and addressing these oppressive behaviors and landscapes and taking immediate efforts to remove these spaces and attitude of privilege (Canva Magic AI, 2024).

During my undergraduate studies, I experienced a lack of belonging in my science program. Fortunately, through the leadership of the institution, special initiatives were created to help students of color engage in campus and academic programs to build a sense of belonging. Participating in these efforts and finding my support network helped me to overcome feelings of isolation and anxiety. Planning our class schedules, studying, and holding each other accountable for doing our best to succeed are some of the strategies that enabled me and my friends to achieve our career goals. Having a space provided me and other African American students to express our frustrations with being ignored and marginalized in our programs of study. When I entered Graduate School, the sense of not belonging returned as many of my peers had other career plans. I was embarking on a new career pathway, science teacher education. I was confident about learning science and carrying out experiments, but now I would be teaching it to others. In my Graduate programs, I was the only African American female in my cohort. The lack of engagement and negative interactions with White faculty and peers presented unique challenges during my academic journey that made me question my ability. When I began to question my ability, I quickly reached out to my family and support network. They reminded me that I have the skills and knowledge to succeed as a science educator and to never forget it.

While in graduate school I observed classroom interactions that tested my confidence and sense of belonging. Instances of my course instructors and peers not demonstrating the practices of high-quality interactions with me, the only African American female or student of color in the from courses were heartbreaking. Even some of my European American classmates limited their interactions with me during course tasks. There were often interactions in classes from the students from U.S. dominant cultured when the instructor allowed students to control class discussions and when I or other students of color offered input in the conversations, they expressed disbelief. They looked shocked that we were able to offer a logical and accurate response to the discussions. Shaver and Moore's (2019) research suggest that most White faculty feel connected more with their White students leaving the African American students and other students of color feeling isolated and not welcomed. My actions of connecting with my support network were a common occurrence during

Graduate School. I also leaned heavily on my parents, mentors, and peers for moral support and encouragement throughout my schooling. I believe these coping strategies evolved and were very effective in helping me complete my advanced degrees. As Haynes (2019) noted I was experiencing many of the unique troubles many African American women in higher education endure as they pursue their degree in fields where they are greatly underrepresented. African American science teachers, especially African American female science teachers were greatly underrepresented in my community and as I learned in my studies they are greatly underrepresented across the United States. Lane et al. (2023) research indicates aspects of how the barriers that African American women encounter impedes their journey to become scholars.

The rest of this chapter will share how I was able to achieve success and cope with feelings of isolation by creating my own positive space for achievement and belonging. My three C's for surviving the challenges of marginalization and preferential behaviors during my schooling: Connections, Communication, and Community. Connections help to improve techniques of active listening and cultivate new relationships to build a system of support (Friar & Eddleston, 2007). Communication is a valuable skill in finding your voice and using it to feel empowered in situations where you have been dismissed. Community empowers people experiencing similar circumstances to support one another through advising and/or mentoring. These three points have enabled me to handle these uncomfortable experiences and achieve success throughout my academic journey. Ellis's (2001) research provides evidence to support the isolation and dissatisfaction African American women encounter in comparison to their male counterparts or White peers. Today, I continue to apply the three Cs in my current role as an African American female science education professor at a PWI. As these institutions of higher education engage in the work of developing more inclusive practices to promote success for women, specifically African American women there remain challenges within the systems. These emerging challenges undermine the accomplishments and abilities of African American women faculty sending the message to students that they do not possess the necessary skills and knowledge of their White counterparts. Haynes (2019) explained that African American faculty, specifically African American women continue to still experience oppression and feelings of marginalization in these institutions. Therefore, the three C's can be applied throughout all of one's academic studies as a practice to foster professional growth and fulfillment. Let's learn more about the three C's and how they helped to shape my resilience and success in academia and the field of science education. Figure 10.1 shows a model of the three C's. The continuous pattern of the model supports its ability to constant movement throughout an individual's academic career and life.

Fig. 10.1 Model of the three C's

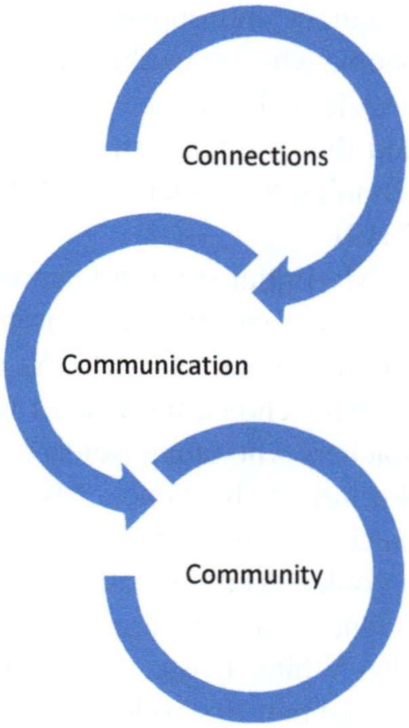

10.2 My Connections

This first aspect, Connections, involves learning how to build and nurture personal interactions. Friar and Eddleston (2007) described the importance and value of this skill to network and gain a better sense of understanding your academic goals. In my AI prompt (OpenAI, 2023), a connection is described as a platform for "building a strong professional network for career success." My connections began early, with the teaching and mentoring of my parents and family members. The nurturing meaningful relationships they provided taught me to appreciate the connections with others who could potentially provide me with the knowledge and resources to succeed. As the only member in my family to aspire to a career in science/science education, building my network of contacts was crucial for degree attainment and professional growth. By establishing new connections with people during my academic journey, I was able to have doors opened for notable job opportunities, collaborations, mentorship, and valuable insights into the field of science education (Friar & Eddleston, 2007).

I recall my parents saying to me numerous times when I was looking for employment. "It is who you know sometimes, that will help lead you to success." I landed my first job as a cashier for my local pharmacy with the connections my mom had with the pharmacist's wife. I applied for the position, and he wrote a stellar recommendation that got me the position! Other examples of having good connections include working as a substitute teacher in my home district, landing my first teaching job and even my position can be credited to my network of contacts. Learning how to make connections and establishing a network of contacts has enhanced my social capital and credibility as a science educator. I currently utilize social media tools, like LinkedIn, to expand my connections in science education research and tap into resources and opportunities to increase my sense of belonging. Connections with people who are working in this field or who are willing to share their experiences are priceless. The connections with these persons better prepared me to handle college and the struggles with finding a space where I felt I belonged and was welcomed (Bradley & Sanders, 2003; Ellis, 2001; Greer and White, 2009; Shavers & Moore, 2019).

My initial support team or network of contacts consisted of my family, friends, teachers, and other educators. Each one played a special role in supporting me to pursue my career goals. As a young child, I did not consider science education as a career choice because of my limited interactions with African American female teachers who taught science. I grew up believing that African American women just did not teach science. My science teachers were White and mostly male. When I entered junior high school, I had my first White female science teacher. She was an older White female, and I was the only African American female student in the science class. In high school, I had my second female science teacher who was also White. It was amazing to learn science from these educators. They played a crucial part in developing my interest in science and created a learning environment that made me feel that I belonged and was valued. While I did not have a desire to become a science teacher at that time, they did provide resources and support to help me succeed in their courses. Interestingly enough the connections I formed with my former high school science teacher helped to land me my first teaching position. She served as my peer mentor! These connections that I made early in my educational studies validated that as an African American female, I needed connections to truly benefit me.

I never really saw myself as a science teacher, let alone a science professor when I began my post-secondary studies. Something clicked when I followed my mom's guidance to work as a substitute teacher during my undergraduate studies. During my semester breaks, I would work as a substitute teacher in the middle schools back home. I was able to build new connections with the teachers and staff employed at the schools and observe how the students interacted with the lessons the teachers planned. I was fortunate to work in science classrooms which allowed me to share my science knowledge with the students. I felt empowered teaching in the classrooms, so I decided to pursue a degree in science education after finishing my undergraduate biology studies. Recognizing how my love for science aligns with teaching science to children landed me in a graduate program to become a science teacher.

Parallel to my undergraduate studies I entered graduate programs where I was one of a few African American female students. Building new connections with people who were finishing the program and were in positions that I wanted to be after graduation. Nelson's (2021) emphasis on the importance of linking up with people who share your "Code" supports the value of connections in attaining the resources and influences to achieve success and satisfaction in life.

By creating connections with other African American women in my science education programs, I was able to create personal spaces to express my anxieties and dissatisfaction with biased classroom behaviors. They were able to provide me with tips on which courses to take to fulfill my degree requirements as well as share details on which professors to consult with for advising and serving on thesis and dissertation committees. My White peers seemed to already have an idea of the courses and processes for achievement but for me, it seemed like a hidden plan that I had to rely on my connections to help me navigate successfully. Starting at a disadvantage is how many African American females feel when we enter these spaces of academia. There is a secret club or handshake that we are not privy to be included. This is the reason why the three C's can aid individuals that are being left on the sideline as the rest of the team enjoys playing the game. Connections and establishing a network of contacts throughout the educational journey can equips you with this information and leverage your ability to excel.

One striking incident that occurred during my methods course helped solidify the significance of having my connections. I was looking for part-time employment as I was completing my graduate studies within the institution. One of my peers connected me with the director of a STEM enrichment program for middle school children. I applied for the position and through my networking skills I was offered the position. Making connections once again landed me an opportunity to thrive in science education. This position as a part-time science instructor inspired my work in creating a STEM, Health care, and Teacher Education enrichment program for middle school youth. It was these connections with my peer and later the director of the STEM enrichment program that helped to shape my approach to science and STEM teaching.

Another example of how my connections have helped to improve my academic journey occurred when I was seeking employment during my doctoral studies. I reached out to my academic advisor which had just received a significant grant researching a local middle school. By communicating with her about my need she was able to offer me a position with the grant and this later led to a post doc position at the institution. Some may consider it to be timing but for me it was my connections that landed me this employment success.

Connections have supported my meeting and achieving academic expectations when situated in academic settings where I am undervalued and expected to fail. My acceptance of this approach has taught me to appreciate my network of contacts, enhance my active listening, and continue to cultivate relationships with individuals who are "dense and diverse" (Friar & Eddleston, 2007).

These connections offered opportunities and scholarships for African American females, like me, to feel included and supported as I continued and completed my academic dreams. Navigating in the biased systems of academia, require connections to promote the academic advancement in overcoming obstacles. These connections continue to evolve and grow through my interactions with new individuals and groups. As an African American female, I have experienced the significant role networking and connections play in allowing me to advocate for myself, fight areas of systemic biases through creating my own spaces where I am included, recognized, and valued for my abilities. The next aspect of reaching my professional goals is communication. You will now read how communication is used in building my connections and providing a community to voice my dissatisfaction with situations that hindered my intellectual and professional growth (Haynes, 2019).

10.3 My Communication

Communication plays a central role in developing social skills with others. It jump-starts the connections among individuals so they can express their thoughts and share ideas. According to ChatGPT (Open AI, 2023), communication is described as a "cornerstone of success." It supports the ability to network and collaborate with others. African American female students often experience situations in academic classrooms where they feel silenced. They feel uncomfortable about expressing themselves when their White peers tend to dominate the classroom interactions without any correction from instructors (Evans-Winters and Esposito, 2010; Haynes, 2019; Robinson, 2013). This behavior signifies the microaggressions, African American students, especially African American women experience as they try to find their voice and sense of identity in these institutions of higher education. Incidents of differential treatment toward individuals from diverse ethnicities and genders is a behavior of racial and gender microaggression (Parson, 2017).

This behavior of silencing is not uncommon in the science classroom. In my schooling, I recall numerous occasions when my White male science teachers allowed my counterparts and White peers to control the class discussions. Suppressing my opportunity to engage in class conversations. As Parson (2017) notes this form of micro aggressive behavior towards African American students, especially females lead to increased anxiety and duress, resulting in lower academic performance and higher attrition rates. After a conversation with my mom about feeling oppressed in the class, she shared her tip of saying, "Open up your ears and listen for the right time, then open up your mouth, and speak your truth." For me this was the beginning of my lesson on the second C- Communication. I had already learned active listening techniques while building my connections and network of contacts, but now it was time to use it in speaking my voice to help negotiate effectively, resolve conflicts, and foster rapport.

People who possess effective communication skills often have positive relationships with other people (Yavuz & Guzel, 2020). However, for African American women communication in classrooms is not perceived as positive when they are silenced, or their comments are constantly being dismissed by instructors and European American peers. While the nature of some classroom conversations may appear positive, the process of communication often evolves to a negative message when the personality of the speaker dominates. Therefore, empathy within these academic spaces is needed to establish and foster healthy environments for all students (Celik & Alpan, 2022). Without empathy training, these spaces in academia will continue to silence students of color and intensify the attitudes of White privilege (Shavers & Moore, 2014). White privilege can be described as "an element of a broader system of racial characterization in which White individual is born with advantages and opportunities that are not provided for people of color (Blakemon, 2017, p. 13).

Success in academia requires effective communication. Creating a process that is open and allows for the transmission of messages that are empathic and accurate is one way these classrooms can increase student involvement and engagement (Celik & Alpan, 2022; Johnson, 1999).

As an introvert, I have never been known to be a big talker. My gift is my ability to be a good listener and provide support in leading others to express themselves. In school, I often felt a little nervous about speaking out in whole group discussions, especially where I was the only African American female. With my speech disability in pronouncing certain sounds, I tried to avoid speaking in my classes. I already felt all eyes were on me as the only African American female student but now I would possibly be singled out or ridiculed in front of my peers. I created my academic mask to protect my feelings and to withstand the dismissal of my contributions to the class discussions. I soon realized that this was not a healthy approach, so I enrolled in a communications course to enhance my communication skills and equip me with the right resources to support my academic journey.

Communication is an important part of the 3 C's! While it may continue to be a struggle for me to find my voice I rely on my connections and community to assist me with my communication skills. Recalling the popular television commercials about EF Hutton in the 1980s, "When EF Hutton talks people listen." I continually strive for a similar response. When I speak about my expertise and work in science and STEM education, I want everyone to listen. Communication has provided me with the support and the voice to speak my truth. I am finally getting the attention and the stage in academia that was once excluded from me because of my race and gender. I bring unique perspectives and insights to academic discussions and by finding my voice and using it I was able to enrich the learning experience for all persons in the classroom. As with many African American women pursuing advanced degrees, they can bring unique perspectives to these academic spaces that are predominately White and restricted. By implementing policies and disciplinary actions that address racial and gender biases in theses educational institutions and will set precedence by creating academic campuses that integrate cultural competency into curricula, provide targeted support services, and establish spaces where students from all backgrounds can share their experiences and learn from each other without judgement.

One personal experience, I will share an incident that happened while I was working on my licensure for teaching science education. During my first methods course, we were required to travel to a local high school in rural North Carolina to observe science classroom instruction for two weeks. I was the only African American student in the course. While observing in the honors biology class, which was taught by a middle-aged White female with well over 15 years of teaching experience, I experienced a rather unwelcoming racial response. The class was ethnically diverse with approximately 28 tenth-grade students. One day while observing the class with my peers, the teacher asked one of my White male peers to join her in explaining to the students the impact of cause-and-effect interactions between different organisms and their habitat. He spoke with a low, monotone voice which was hard for me and the class to hear. He explained some of the factors that organisms might have to live successfully in their habitats. "The factors for a good habitat include having enough space, food, water, and shelter." After he finished talking, the students looked confused. One of the students asked these questions, "What does that look like? Can you give us examples?" My peer looked puzzled, shrugged his head, and sat down immediately. I raised my hand to ask the teacher if I could respond, but she dismissed my question and stated that she wanted another student to explain to the class. The other student was a White female. At that moment I immediately felt dismissed and frustrated. I continued to observe as my peer explained examples and share a handout for the lab activity. After the class, the teacher walked to our group and stated there was not enough time for everyone to talk. She thanked my peers for their input in the lesson. She even praised my White male peer for explaining the information and apologized for not preparing him to answer the questions. When our class returned for our regular class meeting on campus I spoke with my instructor about the incident, and he stated, "You should not take it personally, there was simply not enough time for you to speak. You will have other opportunities to interact with the students." After that conversation, I focused more on using my voice to interact with students informally as they worked on the lab activities. I made it my goal to speak with each group of students in the class during the field visits. I provided them with feedback about their work and encouraged them to use their voices as they worked together to complete the lab assignments and during class discussions.

When I finally got my chance to teach a mini lesson to the students, I explained the concepts using real-world examples to explain how variables can impact population growth, such as the duckweed activity they had completed a week prior. As I taught, I allowed students opportunities to ask questions and clarified any of their misconceptions. Unlike my White peers, the teacher interrupted my lesson by suggesting additional examples in the discussions and adding new expectations for the assignment I had planned. As I wrapped up the lesson, I asked the students to share their feedback on my instruction and it was rather positive. They questioned why the teacher had interrupted my lesson since she did not with the other students.

This incident taught me that the unwelcoming behaviors of White privilege can be found in all areas of education. I learned to communicate my feelings and use this coping strategy to teach the students and the teacher that I was qualified to teach this science class!

I felt empowered after reading the students' comments. Reflecting on this incident and numerous other in which I was viewed as invisible or not qualified, I learned to feel more confident in my academic abilities and created my presence and voice. My aim as a science educator is to help others develop their voices and build on their confidence. In my classrooms learning to speak up for myself and engage in conversations was paramount in helping me to become a successful educator. I am now expanding my skill set in communication by presenting my expertise in science and STEM with future science educators in my community. Having confidence and insight to know that voice is important and without it, one faces many challenges to be recognized and heard in a momentous climate of racial, social, and gender conflict. The final facet of the three C's is Community. Creating your own positive space that supports your feelings and validates your intellectual achievements.

10.4 My Community

According to Meriam-Webster (n.d.) online dictionary, community is "a group of people with common characteristics or interests living together within a larger society". I believe that a community consists of the people, spaces, and connections that help inspire and provide an outlet to reconcile feelings of isolation and marginalization. As a scientist, I think about the term, isolation as the removal of organisms or specimens to prevent contamination or hinder growth. This term from the perspective of an African American female pursuing the professorate reminds me of the constant struggles and frustrations of not being included in opportunities and scholarships that my White peers and counterparts took part in.

Community plays a strong role in my values. From my early childhood days, I saw the community as my second home. Engaging with my neighbors and church family was a common practice. These community members were my teachers, family, and friends and they all contributed significantly to my success in the professoriate through their support and mentorship. For many African American females, like myself, the support of the community is what helps you to stay grounded and accomplish your career goals. The community is my way of having my cheering squad when I feel challenged and oppressed. The best community is one that is composed of individuals that are supportive and honest.

I believe my decision to become a science educator would not have been accomplished without my community. Throughout college, my community has evolved and expanded to include new peers and organizations that shared common experiences of isolation and dissatisfaction in their academic programs. Building my community served as an effective approach to increasing my "motivation to continue and complete- my academic studies" (Moreira et al., 2019; Sowell et al., 2015). I no longer had to feel out of place due to the lack of racial diversity in my classes. Establishing my own community allowed me to address my needs for a more inclusive and supportive academic environment (Davis et al., 2004). The supportive services of my community enhanced my coping skills and increased my resilience and academic success (Parson, 2017).

Having a diverse team of individuals to share insight and experiences strengthened how I approached unwelcoming situations that try to oppress my access to equitable opportunities. Establishing connections with other students from marginalized groups and participating in student-led organizations that advocated for change and provided support for all students and African American students on campus commenced the creation of my expanding community. As Nelson (2021) comments it is important to be strategic in establishing your inner circle, "know your front row." Therefore, community selection should always be intentional and purposeful for the groups or individuals that are included. Today, my community consists of additional groups and organizations that support my work and have missions that align with my values.

Overall, the community is indeed essential for achieving success in the profession by creating connections with others who are either in the same space or have a common social value or belief. In my community I want members who strive to achieve goals by serving others; improving the community by displaying respect for the total community regardless of race, gender, economics, ethnicity, sexuality, or disability; and willingness to work hard to bring growth and togetherness for all. I credit all of the members of my supportive community for helping me to achieve my career goal of becoming a science education professional. The support and resources they provided allowed me to pay it forward. Sharing my three C's with others who are facing similar situations of marginalization and feelings of isolation as they pursue careers in science education. Cultivating a community that informs and provides resources and scholarship is pivotal for those struggling to have their "seat at the table." One must remember with the community you must give back. Find a way to build the success of others and inform society as a whole the need for your work.

Achieving success is an honor and a responsibility to help those who will follow. In my career, I have had the pleasure of interacting with many outstanding individuals. Many have a part in the connections and community of my three C's. They help me to strive for achievement in my current work with providing local middle school children learning opportunities during afterschool times. Without these individuals and the opportunities that they shared my career path would not have been obtained. This is why is it is important to establish a community and follow its purpose of helping others to find success in these challenging environments. Haynes (2019) explains how there is a need for support for these high achieving students. They desire a support system that allows them to feel included and welcomed in these spaces of academic excellence. Rather than suffer in silence like myself and many others before me as we faced the challenges of isolation and oppression.

Through a community you are able to enhance your academic skills and may be offered other opportunities to showcase your knowledge. Through the community I am able to provide service and shape those interested in pursuing science and STEM as a career option. Community does not exist without service. An individual must give back not in expectation to gain but in a self-less manner that contributes to the need or call. Experiences early in my college studies with the enrichment STEM program helped in shaping my vision toward community. By establishing a STEM. Health care, and Teacher Education enrichment program at my institution I found my way to give back. Other efforts that can contribute to the community include volunteering, speaking at events or sharing your expertise. All of these efforts not only allow you to engage in the community but can also enhance your reputation and credibility. For many African American women pursuing acceptance and recognition in the professoriate it will more time than their counterparts or White colleagues. Having an effective community can help shorten this time. A well versed and informed community is priceless in providing the motivation and preparation to excel in academic setting where you are generally disconnected. Therefore, community creates access to the educational opportunities by bridging the gap. Through connections and network groups, African American women can be linked with mentors and educational advocates to help provide guidance.

In academia the unique challenges stemming from systemic biases required the support and solidarity of a community committed to equity and inclusion. The community can serve as a powerful force to not only to help African American women to empower themselves to accomplish their educational goals but to also overcome obstacle through collaboration, change and advocacy. Community is framed around service and through collaborative actions, new pathways for achievement can be formed for every person of color that seeks to fulfill their academic potential. Community is designed to make meaningful contributions to society. By empowering African American women to share their experiences and draw strength from collective wisdom then they are equipped with the coping skills to be resilient as they face adversity in their pursuit of academic excellence.

By narrating my experiences in this chapter, I contribute to the support of aspiring educators of color, encouraging them to persist in their dreams despite systemic obstacles that dismantle institutions supports in the areas of diversity, equity, and inclusion. My overall success as a science teacher and STEM educator stands as a testament to the value of overcoming challenges. In addition to using the three C's one may consider the following guidelines to achieve success in academic and life in the professoriate.

1. Establish clear career goals and expectations. Know your connections and community. Importantly you have to communicate your aspirations with your support team so that they can provide you with the right guidance and resources.
2. Accepting a new role as a life-long learner. Using your community and connections to stay up to date on the latest trends and strategies in your field of study. Always be willing to expand your skill set through trainings and professional development opportunities.
3. Serve your community. Focus on how you can contribute to your connections and community. Show them that you value their support by playing a key role in serving them.
4. Authenticity is the key. Be yourself and truthful. Your authentic self will be more appreciated and respected in your connection and community.
5. Be able to adapt. Do not let the rapidly changing landscape frighten you from achieving your goals. Be open to adjust from any insight or feedback that you may receive from your connections and community.

10.5 Discussion and Conclusion

By leveraging connections, communication, and community, you can find success and fulfillment in your career aspirations while contributing to the growth and well-being of others. Women of African ancestry, Native American women, Latinas, Asian and Pacific Islanders women are marginalized or dismissed as they complete their studies in academia should no longer struggle in silence (Shavers & Moore, 2019). These three C's offer a new practice of resilience for African American women who suffer stress and mental distress as they push themselves toward academic excellence (Haynes, 2019). The necessity for bolstering support for African American women in academic institutions, particularly those where the curriculum and perspectives predominantly champion Western Civilization through a white-centric lens, is both a pressing and complex issue. It's evident that historical and systemic disparities in education have not only marginalized but also silenced diverse voices, underlining the importance of fostering an inclusive and equitable academic environment. Our African American female students and other students of color are constantly facing the questioning of their abilities and identity by their White classmates as well as instructors in academic environments that are excluding the historical accomplishments of African Americans and other cultural groups (Belton-Cardoza & Farmer, 2023). When educational content and discussions largely reflect a singular cultural viewpoint, it risks perpetuating a narrow understanding of the world, thereby limiting the potential for a rich, multifaceted educational experience (Belton-Cardoza & Farmer, 2023; Canva Magic AI, 2024). Increased support in the academic spaces is imperative for several reasons. The constant battles of microaggression negatively impacts the psychological mindset of African American women and other persons of color. Many begin to question their own identities and abilities as to if they are even qualified to be in these academic spaces (Belton- Cardoza & Farmer, 2023).

Even after achieving my academic credentials, the systemic barriers continue to influence how I am perceived in predominately White institutions by the students, administrators, and White colleagues (Lane et al., 2023). Therefore, within the professoriate you can expect to encounter many of the same challenges and feeling of marginalization. Just like you may have experienced during your academic studies, you often begin to question yourself and identity. The views of racial stereotypes and the hostile and unwelcoming climate in these institutions compound the frustrations and oppression experienced within these setting. Often as the only African American woman faculty member the racial stereotypes of the department influence how you think you should interact in academic social activities. Questioning yourself each day about ways to make sure fit in or not be attacked because of your difference. Some questions that you may have asked include the following:

> How should I wear my hair as to not to feel singled out? Should I share my thoughts? Should I not engage in discussions about race as to not make others feel uncomfortable? (Belton-Cardoza & Farmer, 2023, p. 298).

These three C's offer not only African American female students a way to cope and stay resilient as they pursue their degree but to also assist African American female faculty in remembering their identity and ability to be suited for these academic institutions. They should never be afraid to express their authentic selves (Belton-Cardoza & Farmer, 2023). In the current climate of academic institutions where diversity, equity and inclusion (DEI) policies are being repealed (Henkel & Jallow, 2024), these three C's can provide the motivation to continue our work.

By sharing some of my life events in this chapter, I hope this offers guidance to others who are pursuing a future as a science educator. Recalling the supportive relationships that enabled me to create a self-mentoring practice strengthened my confidence, communication skills, and voice in this profession. I learned that staying grounded in my values, which were instilled in me as a young child, allows me to use these three pillars in supporting my professional growth and also in helping others in the community. My three C's are continuing to evolve as I work in different circles and meet different people. Connecting with allies in the community to establish my network of contacts helps me to not only feel valued and included but also validates my position in the professoriate.

I hope this chapter will inspire African American women to use the three C's as a practice to transform how to positively cope in spaces that question their presence and academic abilities. I hope others will see the value of the three C's in building a space that is nurturing, and supportive and cultivates relationships that promote success for individuals to thrive holistically in their pursuit of academic and career goals. There is a community and societal level of responsibility beyond the changes needed in our institutions of higher education to support our African American female students and faculty. They need the encouragement from their programs and departments to participate in STEM fields, leadership positions and other areas where they are underrepresented in order to enrich the entire academic environment by fostering diversity, innovation, and social justice. There remains to be a call for action in making sure the academic experience for all persons of color involves the creation an equal platform where all voices can be heard and foster deeper understanding and appreciation of our diverse backgrounds (Morris, 2016).

References

Bartman, C. C. (2015). African American women in higher education: Issues and support strategies. *College Student Affairs Leadership, 2*, 5.

Belton-Cardoza, R., & Farmer, E. S. W. (2023). Race, color, and racialization in graduate education. In E.S.W. Farmer (Ed.), *The black student's pathway to graduate study and beyond: The making of a scholar* (pp. 295–305). Information Age Publishing, Incorporated.

Blakemon, M. T. (2017). *White privilege*. ABDO Publishing Company.

Bradley, C., & Sanders, J. A. L. (2003). Contextual counseling with clients of color: A "sista" intervention for African American female college students. *Journal of College Counseling, 6*, 187–191. https://doi.org/10.1002/j.2161-1882.2003.tb00239.x

Canva Pro Magic AI. (2024). https://www.canva.com/pro/

Çelik, Ö. C., & Alpan, G. (2022). The impact of an effective communication course with enhanced student engagement on communication skills and empathic tendency of preservice teachers. *Educational Process: International Journal, 12*(2), 33–58.

Davis, M., Dias-Bowie, Y., Greenberg, K., Klutekken, G., Pollio, H. R., Thomas, S. P., & Thompson, C. L. (2004). A fly in the buttermilk: Descriptions of university life by successful black undergraduate students in predominantly white southeastern university. *The Journal of Higher Education, 75*(4), 420–445.

Ellis, E. M. (2001). The impact of race and gender on graduate school socialization, satisfaction with doctoral study, and commitment to degree completion. *The Western Journal of Black Studies, 25*, 30–45.

Evans-Winters, V. E., & Esposito, J. (2010). Other people's daughters: Critical race feminism and Black girls' education. *Educational Foundations, 24*, 11–24.

Friar, J. H., & Eddleston, K. A. (2007). Making connections for success: A networking exercise. *Journal of Management Education, 31*(1), 104–127. https://org.liblink.uncw.edu/https://doi.org/10.1177/1052562906286860

Greer, T. M., & White, A. M. (2009). *Women of color groups: Group counseling with African American college women.* Professional Resource Press.

Haynes, C. S. (2019) A loophole of retreat? Predominately White institutions as paradoxical spaces for high achieving African American women. *International Journal of Qualitative Studies in Education, 32*(8), 998–1018. https://doi.org/10.1080/09518398.2019.1635281

Henkel, C. & Jallow, A. (2024). UNC Board of governors repeals DEI policy, embrace "principled neutrality". *NC Newsline.* Retrieved from https://ncnewsline.com/2024/05/23/unc-board-of-governors-repeals-dei-policy-embraces-principled-neutrality

Howard-Vital, M. R. (1989). African-American women in higher education: Struggling to gain identity. *Journal of Black Studies, 20*, 180–191. https://doi.org/10.1177/002193478902000205

Johnson, M. B. (1999). Communication in the classroom: Guides non-classroom (055) *Opinion Papers* (120). https://eric.ed.gov/?id=ED436802

Lane, T. B., Perez, E., Patterson-Stephens, S. M., & Vital, L. (2023). Research as a site of resistance: A phenomenological study of black women in U.S. doctoral education. In E. S. W. Farmer (Ed.), *The black student's pathway to graduate study and beyond: The making of a scholar.* (pp. 79–96). Information Age Publishing, Incorporated.

Merriam-Webster. (n.d.). Community definition & meaning. Retrieved May 22, 2024, from http://www.meriam-webster.com/dictionary/community

Moreira, R. G., Butler-Purry, K., Carter-Sowell, A., Walton, S., Juranek, I. V., Challoo, L., Regisford, G., Coffin, R., & Spaulding, A. (2019). Innovative professional development and community building activity program improves stem URM graduate student experiences. *International Journal of STEM Education, 6*(1). https://doi.org/10.1186/s40594-019-0188-x

Morris, M. (2016). *Pushout: The criminalization of Black girls in school.* The New Press.

Nelson, S. A. (2021). *The woman code: 20 powerful keys to unlock your life.* Health Communications, Inc.

OpenAI. (2023). ChatGPT (Mar 14 version) [Large language Model- Connection, Communication, Community].

Parsons, A. (2017). The effects of microaggression, predominantly white institutions and support service on the academic success of minority students. *Perspective, 9*(9). http://scholar.unh.edu/perspectives/vol19/issl/9

Robinson, S. J. (2013). Spoke tokenism: Black women talking back about graduate school. *Race, Ethnicity & Education, 16*, 155–181. https://doi.org/10.1080/13613324.2011.645567

Shavers, M. C., & Moore, J. L. (2014). Black female voices: Self-presentation strategies in doctoral programs at predominately white institutions. *Journal of College Student Development, 55*(4), 391–407. https://doi.org/10.1353/csd.2014.0040

Shavers, M. C., & Moore, J. L. (2019). The perpetual outsider: Voices of Black women pursuing doctoral degrees at predominantly white institutions. *Journal of Multicultural Counseling & Development, 47*(4), 210–226. https://doi-org.liblink.uncw.edu/https://doi.org/10.1002/jmcd.12154

Sowell, R., Allum, J., & Okahana, H. (2015). *Doctoral initiative on minority attrition and completion.* Council of Graduate Schools.

Yavuz, S., & Guzel, U. (2020). Evaluation of teachers' perception of effective communication skills according to gender. *African Educational Research Journal, 8*(2), 134–138. https://doi.org/10.30918/aerj.82.20.010

Chapter 11
My Journey as an African American Woman at Predominantly White Institution

Melody L. Russell

11.1 Introduction

11.1.1 My Journey

As I reflect on my experiences as an African American woman who is currently a Professor of Science Education, with an Alumni Professorship at a predominantly White institution, it is important to start with how I got to this point. I graduated from a top, and well known, Historically Black College and University (HBCU) with an undergraduate degree in pre-vet/laboratory animal science. My initial plan was to become a veterinarian until I discovered that I loved the laboratory animal research aspect and shifted my career goals to lab animal researcher/scientist. As an undergraduate student I was in the National Institutes of Health (NIH) Minorities in Biomedical Research Support program (MBRS) which further opened the door to opportunities to conduct laboratory animal research. Moreover, I was mentored by top researchers in the field of laboratory animal science at my undergraduate institution.

The MBRS program was supported by National Institute of Health (NIH) and provided support to traditionally underrepresented students interested in biomedical research. I am grateful for this program because it opened my eyes to a whole new world of research and as a result I decided to become a research scientist. It is unfortunate that these programs are currently under attack and facing defunding because it provided me and so many other students with opportunities to better prepare undergraduate students for careers in the biomedical sciences and graduate studies. I also had the opportunity to work in a summer internship as a laboratory assistant alongside

M. L. Russell (✉)
Auburn University, Auburn, AL, USA
e-mail: russeml@auburn.edu

M. M. Atwater and B. A. Garvin (eds.), *Creating and Maintaining a Home in Science Education Research*, Contemporary Trends and Issues in Science Education 65,
https://doi.org/10.1007/978-3-031-99382-4_11

top scholars and graduate students that enhanced my interest in scientific research. During my summer internships I was an undergraduate researcher at a top tier, private medical school. This experience also furthered my interest in laboratory research in microbiology and immunology. The summer internship was also instrumental in my preparation for graduate studies, and it is important to continue providing funding that support providing opportunities for laboratory research experience for high school and undergraduate students to broaden participation in the sciences. All of these experiences helped me to better understand exactly what I wanted to pursue career-wise and expanded my knowledge and expertise in conducting scientific research. Most importantly was the support I received through my undergraduate institution from mentors and professors that were research scientists who helped me navigate the courses and content. Upon graduation from a HBCU I continued my education at a top tier Research I institution in the midwest and attained a degree in the biological sciences as a dual student in the Food Science and Technology program. This experience allowed me to expand my research knowledge, and I gained a newfound interest in food science and biotechnology. On my journey to pursuing a graduate degree in the field of biological sciences, I discovered that my experience as a student at an HBCU and PWI shaped my interest in broadening participation in science for traditionally underrepresented students like me.

Although, I was accepted to the Ph.D. program at the same institution where I completed my M.S. degree, I was offered the opportunity to return to my alma mater (HBCU) and work as a laboratory research assistant in the same lab where I was an undergraduate student researcher. While working in the research lab as serendipity would have it, one day I was talking to an administrator that told me about doctoral degree programs in Science Education and this piqued my interest. I was not aware that there were doctoral programs in Science Education and after further research I discovered that this would be the ideal degree program to have an impact on broadening participation in the sciences. I learned a lot on my journey as an undergraduate student at HBCU and graduate student at a PWI and wanted to share my experiences relative to how to promote interest and engagement in the sciences. As a result, I applied to a top tier, doctoral program in Science Education at a Research I institution in the Southeast and was admitted. It was an interesting transition since my background was in the sciences and I did not have a degree in education, though I worked hard to learn the content and had an excellent mentor and support system at this institution. My mentor and advisor cultivated my passion for investigating strategies for broadening participation in the sciences, as well as diversifying the sciences, through conducting research in Science Education. This helped to ensure that I was well-prepared for a career in academia.

11.1.2 Background

My journey in academia has been shaped by a number of factors that are the primary reasons for my persistence. This chapter will highlight the significant impact that spirituality/faith, family, mentoring, resiliency, and persistence have had on my journey from assistant professor to full professor at a predominantly White institution (PWI) in the South. Research by Harley (2007) described the role and impact of race fatigue. African American faculty are often overextended with responsibilities and undervalued. More specifically, it is important to use a historical lens when discussing the experiences and stories of African American women in higher education at PWIs. Moreover, a lens of intersectionality is also critical when understanding the various factors that impact the experiences of African American women in higher education due to the interconnectedness of race and gender for African American women. When I started as an assistant professor, I was eager and excited to hit the ground running. The feeling of accomplishment from completing your doctoral program and starting my first "real" job is indescribable (though I did work for one year at as a tenure track faculty member at an R1 university, public university in Atlanta, Georgia). One thing I have learned from the academy and something you hear often is that department heads, deans, and even university presidents will come and go, but faculty stand the test of time. After almost 25 years in higher education, I can attest that this is true.

In addition, I have experienced leadership from a number of department heads (I even served as assistant department head), deans, and presidents and when I started as an assistant professor I would never have imagined the impact that leadership can have on your success (or lack thereof) in higher education (particularly in my experiences at a PWI as an African American women). My first position was as an assistant professor (tenure-track) at a university in the heart of downtown Atlanta. After my first year my husband and I accepted positions at a university in Alabama where I joined the Science Education program. Fortunately, I started in the department with a cohort of new hires, and we all formed a close-knit relationship. Building relationships was probably the first step to acclimating and socializing to a new institution. We were all busy teaching and learning the "ropes" as junior faculty and I soon learned that mentoring would play a critical role in my experience at this institution. Though I don't recall ever hearing about officially assigned mentors I did observe many mentoring relationships develop throughout the department. Furthermore, I did not have someone in the department that was a mentor in the truest sense of the word. However, I was fortunate to have strong support from family, and mentorship from my major professor which was the primary reason I was able to gain my sea legs (so to speak) or footing the first few years.

I admit it was often discouraging because I did not have a mentor within my institution though because of the cohort of colleagues that started together it was a bit easier to still maintain relationships within the institution during the earlier part of my career. I will add that the first three years was a "honeymoon" period only to run into the reality that the honeymoon was over after third year review.

Unfortunately, this experience is not unique to junior faculty, and African American faculty. African American faculty are often welcomed with open arms at the PWI though once you reach third year review perceptions of your work and the value of your work may begin to change. I will add that this seemed to happen to many of the colleagues in my cohort. That said only a few of us are still left recruited during the time started when hired there is always new excitement because you bring in a new lens to the institution, however when this newness wanes oftentimes new faculty members feel isolated, and some may even feel betrayed (or duped). I would characterize it as a "bait and switch" so to speak. The challenge becomes how to maintain your momentum relative to your research even though there is a shift in the tone of acceptance. This can also be impacted by multiple changes in leadership at the departmental level (which was one of the primary reasons for my challenges). That said, there were a number of factors that impacted my persistence and resilience as an African American faculty member. I attribute my faith, family, and mentors, as the reasons I was able to persist.

11.1.3 African American Women Faculty at PWIs

I want to be sure that the deficit model is not used as a lens to interpret the experiences outlined so I will make sure to provide context to the experiences through existing literature. Few studies have examined African American women faculty and their experiences at PWIs (Gregory, 2001; Jones et al., 2015; Griffin & Reddick, 2011) and few have investigated their journey of tenure and promotion and how they navigate at PWIs. Institutions of higher education benefit and have benefitted greatly from having diverse representation of administrators, faculty, students, and staff on the campuses across our country (Gurin et al., 2002; Pittman, 2010). Different perspectives and voices, promote intellectual curiosity, and faculty from underrepresented groups (e.g. African American) as faculty at PWIs may be the only interactions students have with students attending PWIs. Students may or may not interact with diverse groups in high school, or even in their communities (particularly depending upon where the student may have grown up). In preparing students to be socially responsible and better serve their communities, learning from others who may come from different backgrounds and have diverse experiences is critical. This demonstrates the value of differences, as well as the value of similarities. Students and colleagues alike often find out that they are more similarities when engaging with each other.

African American women faculty also have been referred to as "maids in the academe" (Harley, 2008). This is attributed to the historical context of the Black women and perceptions often promulgated of Black women as "taking care" of the community. Historically, there have been a number of derogatory stereotypes (e.g. mammies), through media, books, and society (Collins, 2000). These disparaging stereotypes have permeated all aspects of society and higher education is no exception. Moreover, we live in a racialized and genderized society and institutions of higher education, like PWIs only exacerbate this impacting retention of faculty.

African American women faculty members at PWIs describe feelings of anxiety, frustration, and alienation (Frazier, 2011; Griffin, 2012 as cited by Jones et al., 2015). Furthermore, African American women experience a litany of stereotypes based on race and gender often resulting in their being devalued particularly at PWIs (McKay, 1992 as cited in Harley, 2008). Scholars have examined the experiences of African American women faculty at PWIs and the challenges and hardships they may encounter as they pursue tenure and promotion (Alston, 2012; Turner, 2002). These experiences have been further identified as "gendered racism" where there are situations when they are told by their White colleagues that they would be better off at an HBCU (McKay, 1992 as cited in Harley, 2008). Currently, it goes without saying that the vast majority of faculty at PWIs are White males, primarily at the rank of full professor. African American women have been in various roles at PWIs for over century. Subsequently, from a historical standpoint African Americans at PWIs were largely in service-related roles or held specific appointments (e.g. Dean of women), in the case that they were employed by PWI's (Harley, 2008). Furthermore, research demonstrates that African American women have not integrated into the culture of PWIs to the extent of their White counterparts and receive far fewer opportunities for collaborations on research compared to White women (Alfred 2001 as cited in Harley, 2008; Turner & Myers, 2000 as cited in Harley, 2008; Turner, 2002).

Based on research by Harley (2008) African American women at PWIs are often expected to wear several hats and in many cases, they are over-worked and under-valued which results in a type of very real "race fatigue". Harley (2008) even characterizes the experience as the "negro in residence", though I go a step further and have heard my African American colleagues in general say that we are collectively perceived as the "negro of last resort", where we are rarely involved at the beginning, though always contacted when the problem often beyond repair. Subsequently, after all options are exhausted, we are still expected to be the "fixer" or "clean-up person". Furthermore, when we resist this role (since we were not consulted earlier on), and our opinions weren't valued initially, then we are often targeted as not being a "team player", or lacking collegiality. Moreover, we are rarely contacted when there are triumphs, just when there are trials and tribulations which is also emotionally taxing for African American faculty. I have talked to African American faculty that would love to hear from students not just when there is a crisis, but when there is something to celebrate. To take this a step further, we are rarely acknowledged for our advocacy though our White counterparts are acknowledged and considered the heroes for even the most minimal support.

Harley (2008) highlights several aspects of the African American woman's experience in PWIs which I will recount since it resonates in many ways with my own experience. These include, (a) the role of racism in academe, (b) construction of profiles in their role, (c) coping strategies and psychosocial aspects, (d) spiritual and physical aspects, and (e) legal aspects. It goes without saying that I as an African American woman have also experienced these factors on my journey from assistant professor to full professor.

11.1.4 Racism and the Academe for African American Women

Racism manifests itself in numerous ways in the academy. According to research by Harris (2007) African American women faculty experience more disrespect from students than their White counterparts, and more unique challenges (e.g. marginalization, academic bullying). However, it is often difficult to make the experiences tangible. For some African American women, they may feel undue pressure to "fit-in" or compromise their own beliefs and authenticity (e.g. behaviors, attitudes), for fear of being stereotyped as the "angry Black woman, (Jones et al., 2015). According to Frazier (2011), the stress and academic bullying can have debilitating effects on African American women faculty, which can adversely impact their success (e.g. tenure). There is no doubt that being successful (often characterized by tenure and promotion) in academia is a challenge and highly competitive for all, although the types and levels of challenges differ across racial, ethnic (Al Ariss et al., 2014) and gender dynamics (Silander et al., 2013). Institutions of higher education and PWIs specifically were not made for African Americans, and though the climate on campuses has improved, racism is "baked in the cake" in so many institutions and it will manifest itself when the majority feels threatened (the current political climate is a testament to this). Though efforts to diversify higher education and promote equity is nowhere near perfect, it is still a work in progress and necessary to promote equity in higher education, so college campuses look more like the communities they often serve. Racism, often manifested as microaggressions (DeCuir-Gunby et al., 2020; Pittman, 2010, 2012) and macroaggressions (Awad et al., 2021; Sue & Spanierman, 2020), can be from colleagues, administrators, and even students through various encounters. Moreover, some of these experiences raise to the level of macroaggressions. As highlighted in Harley (2008) saying that racism is a cause to a particular effect is often received as inflammatory and though the process may appear unbiased the outcomes are often difficult to explain. Subsequently, community is key for African American women faculty to navigate these experiences, and central to this community is their faith and spirituality, family, mentors, and colleagues who are advocates. African American women faculty also learn to manage the stressors of academe thorough networking and sharing their challenges (Harley, 2008).

More specifically, when an experience or encounter can't be explained it is often said that the experience can usually be placed into a category of one of the "isms". The tenure and promotion process I experienced was marred by a number of mixed messages from colleagues and leadership. This put me in a bit of a conundrum as to who to listen to and experiences like this often alienate and isolate candidates as they work towards tenure and promotion. Unfortunately, colleagues even go as far as to recommend tenure for a candidate without promoting them, sending the message that you can stay, though you are not good enough to be considered equal and receive promotion. This reiterates Harley's (2008) assertion of African American women being considered "maids in the academe", metaphorically speaking. Moreover, colleagues may not respect or value your voice, and the colleagues that are the most challenging to build relationships within my experience at PWIs are White women. The tenuous relationship between African American women and White women throughout history from the enslavement of African Americans, through Jim Crow, has been longstanding. I have encountered hostility and suspicion from some White women faculty members (though I have also encountered White women faculty that were supportive). Unfortunately, African American woman in academe, are often being placed in situations intentionally designed to see if they could prove the value of their scholarship and justify their presence at the institution. What was always ironic about this, was in the not-so-distant past White women did not have the rights to attend the same institution. One example of cynicism and hostility was when I was awarded a research grant that I was added to because a colleague left. I remember the excitement of receiving this grant and sharing this news, and the response I received was "why would they add you to this project"? as if to say what qualified me for the project. Simply put, some White women in my experience have become the problem that they fought so long against, and they essentially have become the "gatekeepers"" now that they are in position of authority or power since there are more women in the field of education.

Other examples are when I have learned from students and even colleagues that I was disparaged in conversations by White women colleagues, and I even had colleagues and students share how uncomfortable they were with the conversations. Of course, this is extremely unprofessional, though the nature of the beast at PWIs. Finally, one example that probably resonates with many African American faculty are the teaching evaluations. Unfortunately, there are instances when these evaluations may be really more personal and not reflect your teaching, and this can be used to adversely impact your merit by department heads if they are not advocates. Fortunately, more recently, department heads and leadership at institutions of higher education recognize the biases in teaching evaluations, so it is important to take these with a "grain of salt". Take the information that is valuable and addresses your instruction and leave the stuff that is really just a personal attack if there are students that don't like you. I can even recall many years ago in a teaching evaluation where a student complained that I ordered pizza for a class, and she didn't want pizza (or something along those lines). That's why I also learned to take some good advice from mentors and colleagues and not read evaluations right after the course. Wait a while after the class has passed and then take a look at them. The more time passes the easier they are to read.

Coping strategies for the challenges encountered on my journey centered around my faith and spirituality, family, mentors, and advocates (some of whom were White colleagues and administrators were the primary reasons for my persistence. Moreover, resiliency has been a trait that has helped motivate me to continue at a PWI. In addition, to resiliency and being intrinsically motivated to persist, I always remember my mentor saying how important it is for us to remain in these spaces. For me it is important to remain at a PWI because if the few African Americans leave, then who will be the voice and face of African Americans and African American women at institutions of higher education. We were left out for way too long so we cannot turn back. The research we conduct is transformative and necessary so we must continue to be a voice for ourselves at PWIs, otherwise someone else will speak for us and then our stories are no longer authentic.

11.1.5 Faith and Spirituality

Faith and spirituality have been the hallmark of African and African American culture for generations. Our ancestors relied on their faith through heinous experiences whether that be the journey as enslaved people from the Motherland or working tirelessly as the backbone of this country (even helping to build the White House) while being treated as less than human and having it justified through White supremacist ideology (e.g. Jim Crow). The Black church has always served as a cornerstone for the Black community and my strong faith in God has been the glue that has held me together coupled with my family support system guiding me through the various challenges. More specifically, I have faced challenges with tenure and promotion from assistant to associate professor, even some challenges when applying for promotion to full professor. Even when I felt like I was "hitting my stride" there would be another obstacle, and I would rely on my faith to help sustain me throughout the very challenging times. In addition, my family circle (husband, son, parents, sister, extended family, and mentors) all helped encourage me to rely on my faith. Though it was also stressed to do be sure to do the work (always knowing that we as African Americans have to exceed the standard) while keeping the faith and staying in prayer. A look back and history shows that African American culture in the United States is steeped in a strong sense of faith and belief in a higher power. This belief and religion have been used to help us communicate with and deal with stress, etc. From then, to present day, faith and spirituality are closely aligned with our psychological makeup, as well which is one of the factors that has helped African Americans persevere in spite of adversities and challenges. This faith also helps to celebrate victories and is a testament to the reward after being steadfast in faith. My faith is definitely a key factor in sustaining me through challenges and keeping me positive and thankful for triumphs.

Moreover, my faith (which goes being just religion) is aligned throughout my physical, mental, and spiritual well-being. My faith and spirituality have also helped me to stay grounded and emphasizes the importance of "paying it forward" and supporting other colleagues in their journey when they encounter similar challenges. Finally, strong faith has helped me to always believe I can do things that may seem impossible and also helped me maintain my self-confidence. This is extremely important when African American women may experience situations where their colleagues (from various racial/ethnic backgrounds) may make them feel insignificant or "invisible" at PWIs and sustaining faith in your own self-worth is critical when you are in the "minority" at an institution. This faith also ties in directly to resiliency and persistence which I will also address in this chapter.

11.1.6 Family

In addition to my faith, my family has played an important role in my persistence at a PWI as an African American woman. My family (husband, son, parents, sibling, uncles, aunties, grandparents, the list goes on) have always been my biggest cheerleaders. As a first generation in my household and one of the few in my immediate family that pursued an advanced degree, they have always supported my endeavors in academia. When there were challenges, I always had someone to talk to and vent my frustrations and it was always an empowering space with my family. They also helped me to stay the course, and rely on my faith, to continue in my goals of tenure and promotion. Simply put, when the going got tough, I was always encouraged to get tougher and know that God put me in this place for a reason and my success would be a testimony of faith, perseverance, and belief that I could do all things through my faith in Christ.

I grew up in a middle-income, working-class household, with a very close-knit family of support, and strong work ethic. I had amazing family network (e.g. grandparents, uncles, aunties, cousins, and family friends) and they were always encouraging me and rooting for me and my little sister growing up. This support was also key to my resilience when I encountered challenges. Moreover, it is important in our culture to make our family proud and not be a disappointment which was another reason why I carried this ethos throughout my experiences in higher education. Unfortunately, it is well known that African Americans are judged as a group by their actions and not individuals so it's important to always be aware that what you do can often reflect on the culture (particularly, when not successful). For example, as an African American faculty member if for some reason you don't achieve tenure and promotion or "fit-in" colleagues may be more likely to attest this to us not being prepared or able to make it and attribute our lack of success in academe to being African American. This sadly, enough may remain in the back of the minds of faculty that may through unconscious biases think twice before hiring another African American faculty member.

11.1.7 Resiliency

Resiliency for African American women in higher education is one of the cornerstones of coping and persisting both personally and professionally (Harley, 2008). Resilience theory is guided by the ability of an individual to persevere and do well in spite of adversity and challenges (Harley, 2008). The ability to manage stress, adapt, adjust, and continue in spite of challenges is key to persistence in academia, and this is even more necessary at PWIs. Key to this how African American women faculty network and find support either within, or outside their institution (Behar-Horenstein et al., 2012). There are a number of examples of where I have experienced resiliency right after though a specific example comes to mind. In keeping with the theme of this chapter my promotion from assistant professor to associate professor is the most apparent example. There were a lot of challenges to more tenure and promotion, and had it not been for my support system and faith I would probably not still be a faculty member at my current institution.

11.1.8 The Role of Mentors

Mentoring plays a critical role in the persistence and resilience of faculty in academia. This is also very important for the persistence of African American faculty at PWIs. Mentors can be from within the institution or beyond and may occur formally, informally or even emerge organically based on faculty that may bond based on similar interest. Research has demonstrated the key role mentoring plays in adjustment and socialization of faculty from culturally diverse backgrounds (Bradley, 2005; Jones et al., 2015). There are a number of mentoring models that institutions use for new faculty (if they use a model at all). When I started at my institution as far as I can recall we did not have a formal mentoring program. We finally instituted a formal policy in my department, and I assisted with this process as part of a leadership team when I served as assistant department head. Unfortunately, depending upon the culture of an institution (and the college and department) the onus is on the junior faculty member to reach out to a senior faculty member, if there are no formal policies to ensure equity and support for all new faculty. It is also important to note that even faculty that have been tenured and promoted to associate professor still need intentional meetings.

As a new faculty member some institutions provide formal mentoring programs, pairing them with more senior faculty. This may be one-on-one mentoring or may include a cohort of mentors. Unfortunately, the challenge of formal mentoring programs is that the faculty may not have similar interest or styles for engaging. Depending upon how big the program is a new faculty member may have to be mentored by a colleague that is not in their discipline or may not use similar research methodologies. This can also cause a mismatch and frustration for both faculty members. There are also some instances where there is little communication relative to the expectations of the mentor or mentee. I often say fictitiously to graduate students that "we all remember all the courses on mentoring we took in graduate school". They always look surprised when they realize that I am joking and for the most part faculty do not get any experience with how to mentor (relative to their graduate students or colleagues). Another source of frustration for junior faculty that I have noticed is when some junior faculty are mentored by faculty that they may have already had a relationship with when they arrived at the institution. Subsequently, this can result in inequities relative to opportunities and access to experiences that better prepare a new faculty for tenure and promotion. Research by Davidson and Foster-Johnson (2001) demonstrates multiple components of mentoring. These components are based on both psychological and emotional support, career and professional development, as well as role modeling (Bradley, 2005; Davidson &

Foster-Johnson, 2001). Some researchers go a bit further and describe mentoring as power-sharing, information-sharing, and self-sharing (Crawford & Smith, 2005).

Mentoring plays a strong role in mitigating barriers to promotion and tenure from faculty from culturally diverse backgrounds (Gay, 2004). Furthermore, faculty that are mentored have reported enhanced job satisfaction, promotions, and increased salaries (Johnson-Bailey & Cervero, 2004). This highlights the importance of intentional mentoring to persistence and perseverance in academia. Subsequently, faculty that lack a mentor (or mentors) may be adversely impacted when applying for tenure or promotion and are more likely to leave the institutions. Depending upon whether or not the mentor is in the department or an administrator/supervisor for a junior faculty member can also have a significant impact on their advancement.

Mentors in the department that are advocates (particularly) if they are senior faculty may have a certain stature and level of respect and their support of a faculty member applying for tenure and promotion can have a significant impact on other colleagues in the department. This is also true if the mentor is in an administrator role in the department. However, it is important to highlight that this can also have an opposite effect on a junior faculty member that is striving for tenure and promotion. Subsequently, if the mentor and mentee have a relationship that is not productive, then this can also provide an undue influence in proceedings and meetings that surround the junior faculty members' application for tenure and promotion. We all hope that people are fair and don't have person vendetta's though institutions of higher education do not operate in a vacuum and personal opinions, biases, discrimination, and prejudices (coupled with stereotypes) can have a detrimental effect on a junior faculty members goal for tenure and promotion.

In addition, the presence or absence of mentors that are advocates can also have an impact on how junior faculty members are viewed in their own discipline (if their mentor is in their discipline). Oftentimes, mentors that are advocates recommend their mentees for awards and special recognition. In the absence of a mentor this can impact opportunities for various honors and awards which are also important for building a strong curriculum vitae. In my experience African Americans and African American women have rarely been recommended for awards. This may be because they may not have close mentoring relationships with so called "powerbrokers" in the department, college, or institution and may often be overlooked. More specifically the work that African American women do is often undervalued (e.g. mentoring, advising, service, committee work, administration) and their work is often called "invisible labor", which is very rewarding for the faculty member, though rarely recognized in tenure and promotion considerations.

Furthermore, the invisible labor, is extremely time-consuming and interferes with the time that need to be spent on research. This is another reason why mentors that serve as advocates are necessary because they are able to negotiate with administrators on the behalf of the junior faculty member and may be able to help them navigate and learn when to say no, and what opportunities are important to pursue. When African American women faculty are asked to serve on committees, they are often the only person from a diverse background at the PWI, and this can also place undue burden to agree since there may be no diverse representation if they decline.

I have observed that faculty with mentors that were advocates were shielded from work that did not enhance their research productivity or were supported if they did have additional responsibilities that may have detracted from their research (e.g. course release, graduate student assistants).

I have experienced a number of mentoring models and mentors outside of my institution have been the most effective mentors. I have been fortunate in that my mentoring network has supported me and provided knowledge and encouragement for navigating some of the most challenging situations. My strongest mentors have been African American faculty (women and men), and they also challenged me the most. They always celebrate my triumphs and have always encouraged me to continue my journey, which is one of the primary reasons I remained at a PWI. I will also add that cross-cultural mentoring, is another factor that occurs as African American women faculty navigate PWIs. I have also had support from White men who were in leadership positions at senior levels at the institution. Typically, this phenomenon may be because they are not threatened "so-to-speak" at PWIs since these institutions were essentially built for them. Subsequently, when they are at the senior level of their career, they are more open to sharing their knowledge and experiences. Not to say more junior level faculty are not able to mentor, I have just observed that they may be busy with their own career juggling and balancing their research, teaching, etc. and working towards tenure and promotion too.

I would also say that mentoring by White women faculty members has been rare or non-existent and this may be because they are already at a marginalized position in the institution as women so navigating while mentoring and preparing for tenure and promotion may be just as much of a challenge for them, when considering the patriarchal structure (e.g. White male dominated) of PWIs. As it relates to cross-cultural mentoring, I have observed that mentors at the senior level, with a significant amount of institutional knowledge seem to be more eager and experienced to mentor and advocate. Finally, I would be remised if I didn't caution African American women faculty to not assume that other women faculty at their institution are open,willing, or have the time to serve as a mentor. I have also learned that networking among African American women faculty must be intentional, and it is important that African American women faculty at the junior level seek out and network with other African American women faculty. In addition, I have learned in my years in the academy that the great literary poet and writer Zora Neale Hurston was correct when she said, "Your skin folk ain't always your kinfolk".

I will also highlight that it is important to provide intentional mentoring at the associate level too. I have seen several situations where once a faculty member is promoted to associate professor they become "stuck" in the role of associate professor, and it is difficult to reach the professor level. Promotion to full professor can be particularly challenging for African American faculty and women due to "invisible labor" for the tenured and promoted associate professor. Additional responsibilities on committees, (e.g. chair), increased teaching loads, increased advising and mentoring loads, administrative positions, and service can impact research productivity. Subsequently, without intentional mentorship at the associate level (e.g. advocates) for support and encouragement relative to research productivity (and learning when to say no to various roles and responsibilities) faculty can become overburdened and this can hinder progress towards promotion. I have also learned that the culture of the department plays a key role in who is promoted to full professor.

11.2 Conclusion

In conclusion, I have given an overview of my experiences and an African American woman at a PWI and highlighted research that aligns with my experiences. The experiences I outlined in this chapter and my faith, family, mentors have served as a support network and these factors are the reason I can call the PWI I have worked at for over 20 years, home. Though I have had some challenges, I have had more triumphs, and I am grateful for the opportunity to serve as a faculty member at a Research I institution. In addition, I have had the opportunity to attend and work at a number of institutions, and I have also been a student at multiple institutions (e.g. HBCUs, PWIs) and experiencing different settings did help prepare me for the lack of diversity in among the faculty and student body at PWIs. Hopefully, my presence at the PWIs has provided an insight and promoted students to engage more in experiences that teach them about themselves and the importance of culturally diverse faculty from an asset model and not a deficit model. Though I have experienced many challenges I have also experienced a number of triumphs support, encouragement, and made dear friends both within the institution and in the local community for all different backgrounds and walks of life.

I have always been true to myself and shared my voice and even though it may have been the reason for my not always being accepted, I know that I have always maintained my integrity and stood by my opinions (even when challenged) or when there were consequences that could adversely impact my career. I am not a "go along to get along" kind of person and in academia as a person that is not in the majority this can have longstanding consequences relative to career goals. My faith, family, mentors and a network of support has helped me to navigate and make the best decisions for my career and maintain my sanity when faced with conflict and challenges that oftentimes had little or nothing to do with my performance or ability. My primary goal as one of the very few African American women senior faculty members at a PWI is to be able to "pay it forward" and provide insight to junior faculty and students (undergraduate or graduate level) to persevere and accomplish (and even exceed) their goals. I know that I would not have the opportunity to work at a PWI had it not been for the sacrifices of the ancestors and my goal is to continue to work hard, be a voice for justice and equity, and hopefully break down barriers to success and pursuit of tenure and promotion by other African American women faculty at PWIs. Unfortunately, we still live in a society where a number of myths and stereotypes about African Americans, African American women exist across the world. I know that as I do the work that I do at a PWI (e.g. teaching, advising, service, outreach) that I am doing "good work" and getting in to "good trouble" in the words of the Late Congressman John Lewis. I realize that I am standing on the shoulders of giants and hope that I make the ancestors proud and continue to persist, and thrive. I do not take the opportunities that I have for granted, because I know that what I have achieved is no small feat and I am my ancestors' wildest dreams.

11.2.1 Recommendations

One recommendation I would give newer faculty, and graduate students seeking a position in higher education is to either have a good relationship with your previous advisors or be intentional in finding a mentor that is doing what you want to do and stay connected with them. This will be your go-to person when seeking counsel on decisions as often as possible. When applying for positions if you are contacted for an interview be sure to find out who the faculty of color are in your program (or even someone in a different college) and try to meet with them during the interview for dinner or lunch. If you are interested in a position at a PWI find out if there is an organization for Black faculty and try meet with them to get a pulse on the climate at the institution for advancing Black faculty. Unfortunately, you may not end up with a strong mentor that is a senior faculty member within your institution so it is important to be sure to stay connected to your major professor or attend conferences in your field to meet faculty and administrators that can provide advice and support.

In addition, find someone in your department that is a senior faculty member that you can build a rapport with early on. This should be an authentic relationship, and it may take a while to develop a rapport with this faculty member, though they may end up being your biggest advocate. It is important to have someone in your department that is well-respected and has some influence with other faculty. The only caveat is that this individual may not be interested in being an advocate, though time will tell as you get to know more faculty members in your program and department. This may even be someone that is promoted to full professor that you built a relationship with prior to their promotion.

Also, be sure that you have a good rapport (if possible, with your leadership). I have had a number of department heads and Deans, and it was always important to have someone to go to when there were challenges or issues. Moreover, when or if you have an issue with a colleague, or even your department head, it is important to make sure you have kept good records, and always make sure you read the faculty and department level handbook, guidelines, etc. and follow the protocol outlined in the faculty handbook for everything from tenure and promotion, etc. One good piece of advice I like to offer is when you are faced with an issue always determine if this will be a "hill worth dying on". Subsequently, don't make a "mountain out of a molehill", and make sure you got the receipts for anything that you do know is an issue before taking it to the leadership. This is especially important if you are a junior faculty member or even at the associate level. That is another reason why it's important to have a senior level colleague in your department that is an advocate in case there are situations that you run into that may impact your opportunities for tenure and promotion. In a perfect world, people would not hold grudges, or biases, though we know we don't live in a perfect world and institutions of higher education do not exist in a vacuum. This is evident given today's hostility and the political climate towards institutions of higher education. Moreover, this person may also be able to offer advice and insight on various situations since they have been in the department for a while.

One thing I learned much later was the importance of negotiating and changing your allocation. This is important to help ensure that the time you spend on research, teaching, outreach and service is counted accurately towards your progress and doesn't end up hurting you in the long run. For example, if you spend a significant amount of time on outreach then be sure to negotiate with your department head an allocation that adequately reflects the time you spend in outreach. Also, be careful not to commit to too much service as a junior faculty member and a senior level faculty member or your department head can help ensure this doesn't happen.

I would definitely recommend you have a hobby or something to do for stress relief too. Higher education is very stressful both mentally and emotionally so it's important to take time for yourself, spend time with family, friends, and loved ones, and always make sure you schedule breaks for your own mental health, emotional health and well-being. At odds with trying to do something in your spare time is the fact that there is never a lot of spare time so it's important to manage your time and schedule so you can do things outside of academe. I picked up hobbies of making jewelry, crocheting, sewing, and I love to travel with my family.

The environment at PWIs can be a challenge, though sometimes more than others. There are great perks to being in the academia as a professor and you will have a number of opportunities and experiences that will help you grow and become a better person. The opportunity to touch so many lives of students, engage with so many different colleagues, and travel to so many places to share your passion for your research and hopefully interest others in your scholarship, while learning from so many different scholars makes this career very rewarding. That said, there will be times that will test your patience and make you question if this was the right choice. However, at the end of the day you know that your work and presence has made a difference in so many ways, even if it is just the mere fact that many people you have encountered along your journey in academia may have never learned the true value and importance of diversity in higher education had you not been a part of their institution and community.

I will also add that I learned to be very careful of who you trust as a colleague relative to collaborations. Though you may have a sincere interest in collaborating to help your colleague and further research in your area, there are some that may take advantage of your kindness and end up stealing projects and ideas from you, trying to make you do all the work while they try to take credit for work, and even worse, diverting potential projects and assignments from you without your knowledge for themselves. Always keep your ideas, your talents, and projects as close to your vest as possible unless there is a colleague that you know based on personal experience (or you have learned from other trusted colleagues) that the potential collaborator can be trusted. In summation, these experiences are my own and others may not have had these experiences. Moreover, strategies that helped me cope and be successful at PWIs may not be a "one size fits all". It is important to find out what works for you since your story and your experience are unique to you.

References

Al Ariss, A., Özbilgin, M., Tatli, A., et al. (2014). Tackling whiteness in organizations and management. *Journal of Managerial Psychology, 29*(4), 362–369.

Alston, J. (2012). Standing on the promises: A new generation of Black women scholars in educational leadership and beyond. *International Journal of Qualitative Studies in Education, 25*(1), 127–129.

Awad, M. N., Crusto, C. A., & Hooper, L. M. (2021). Macrointervention processes and strategies for leaders, changemakers, advocates, allies, and targets: A new framework to address macroaggressions in systems. *New Ideas in Psychology, 62*, Article 100858.

Behar-Horenstein, W.-O., Moore, H., & Roberts, K. (2012). Resilience post tenure: The experience of an African American woman in a PWI. *Florida Journal of Educational Administration & Policy, 5*(2), 68–84.

Bradley, C. (2005). The career experiences of African American women faculty: Implications for counselor education programs. *College Student Journal, 39*, 518–527.

Collins, P. H. (2000). Gender, black feminism, and black political economy. *The annals of the American academy of political and social science, 568*(1), 41–53.

Crawford, K., & Smith, D. (2005). The we and the us: Mentoring African American women. *Journal of Black Studies, 36*(1), 52–67.

Davidson, M. N., & Foster-Johnson, L. (2001). Mentoring in the preparation of graduate researchers of color. *Review of Educational Research, 71*(4), 549–574.

DeCuir-Gunby, J. T., Johnson, O. T., Womble Edwards, C., McCoy, W. N., & White, A. M. (2020). African American professionals in higher education: Experiencing and coping with racial microaggressions. *Race Ethnicity and Education, 23*(4), 492–508.

Frazier, K. (2011). Academic bullying: A barrier to tenure and promotion for African American faculty. *Florida Journal of Educational Administration & Policy, 5*(1), 1–13.

Gay, G. (2004). Navigating marginality en route to the professoriate: Graduate students of color learning and living in academia. *International Journal of Qualitative Studies in Education, 17*, 265–288.

Gregory, S. T. (2001). Black faculty women in the academy: History, status, and future. *Journal of Negro Education*, 124–138.

Griffin, K., & Reddick, R. (2011). Surveillance and sacrifice: Gender differences in the mentoring patterns of Black professors at predominantly white research universities. *American Educational Research Journal, 48*(5), 1032–1057.

Gurin, P., Dey, E., Hurtado, S., & Gurin, G. (2002). Diversity and higher education: Theory and impact on educational outcomes. *Harvard Educational Review, 72*(3), 330–367.

Harley, D. (2008). Maids of academe: African American women faculty at predominately white institutions. *Journal of African American Studies, 12*, 19–36.

Harris, T. (2007). Black feminist thought and cultural contracts: Understanding the intersection and negotiation of racial, gendered, and professional identities in the academy. *New Direction for Teaching and Learning, 110*, 55–64.

Johnson-Bailey, J., & Cervero, R. M. (2004). Mentoring in black and white: The intricacies of cross-cultural mentoring. *Mentoring & tutoring: Partnership in learning, 12*(1), 7–21.

Jones, B., Hwang, E., & Bustamante, R. M. (2015). African American female professors 'strategies for successful attainment of tenure and promotion at predominately white institutions: It can happen. *Education, Citizenship and Social Justice, 10*(2), 133–151.

McKay, N. (1992). A troubled peace: Black women in the halls of the white academy. *The Bucknell Review, 36*(2), 21.
Pittman, C. T. (2010). Exploring how African American faculty cope with classroom racial stressors. *The Journal of Negro Education, 79*(1), 66–78.
Pittman, C. T. (2012). Racial microaggressions: The narratives of African American faculty at a predominantly white university. *Journal of Negro Education, 81*(1), 82–92.
Silander, C., Haake, U., & Lindberg, L. (2013). The different worlds of academia: A horizontal analysis of gender equality in Swedish higher education. *Higher Education, 66*(2), 173–188.
Turner, C. (2002). Women of color in academe: Living with multiple marginalities. *The Journal of Higher Education, 73*(1), 74–93.
Sue, D. W., & Spanierman, L. (2020). *Microaggressions in everyday life*. Wiley.

Chapter 12
Found a Home: Freedom Scie-Cycling

Noemi Waight

I found home outside of the harmful walls of the academy; I found home in my extended community and mentors, my biking community, my supportive family, and closest friends, and in the depths of my familial and ancestral lineage. Most importantly, I found home on my bike, in the depths of my soul, feeling free and being one with the wind and nature, experiencing joy. In this chapter, I map the journey of finding and naming a scie-cycling home—an activity and space—a counterspace—that is the present and stands at the crux of past experiences and a re-envisioned future. Finding a home and a science education home involved circumventing the "metaphorical spikes" (Benjamin, 2019), the intentional and unintentional efforts to systemically exclude and marginalize women of color in the academy. This necessitated a return to the young, curious, vociferous reader and full-of-life Black girl who grew up on Garcia Street, playing with worms and grasshoppers and keenly observing them regenerate and hop, respectively. It involved re-centering a dream, anchoring joy, and forging a commitment to decolonize and strip the spikes, to create a "freeing" counter-pathway that can serve as a framing for the next generation of science education scholars of color.

For most women of color, finding home in the academy—feeling safe, valued, supported, and with opportunities to thrive is not generally the narrative that frames their experiences. In contrast, "women of color face harsh realities in their professional lives as university faculty members. At the same time, even within the walls of [and outside of] these often-pernicious academic environments, women of color can assert their voices, effect change, find allies, and not only survive, but thrive" (Niemann, 2012, p. 446). In the analysis of the countless stories told by women of color in the academy, Niemann states that these stories "demand a comprehensive theory of the foundations of tokenism and related damaging experiences too often

N. Waight (✉)
University at Buffalo, Buffalo, USA
e-mail: nwaight@buffalo.edu

M. M. Atwater and B. A. Garvin (eds.), *Creating and Maintaining a Home in Science Education Research*, Contemporary Trends and Issues in Science Education 65,
https://doi.org/10.1007/978-3-031-99382-4_12

sustained by members of underrepresented groups in these places that are supposed to be bastions of knowledge and enlightenment" (p. 447).

The literature is replete with documentation of women of color who are constantly challenged, retaliated against, gatekept, and suffocated (see Niemann et al., 2012). At the same time, there has been a lineage of scholars of color who, despite harmful experiences, have been successful in the system (Niemann, 2012). In one example, A. Susan Jurow, in her talk, "Can We Design for Healing in the Learning Sciences?" at the 2024 Annual Meeting for the International Society of the Learning Sciences (ISLS), shared her story of institutional harm and its impact on her scholarship and mental health; as a mechanism to counter these harms, she offered guidance on healing through principles of communal action, engagement with the creative arts, reimagining methodology and evidence, and embodying care in research and pedagogy (CARE). These stories of harm, which are the norm in higher education, are often only anecdotally shared within Black and Brown affinity spaces. Indeed, there is a notable gap and absence of narratives of scholars of color in academia, and speifically in science education—that despite institutional harms, cultivate counter-spaces to thrive, dream, imagine possibilities, and center joy. This chapter seeks to address this gap by outlining a Freedom Scie-cycling framework (FScF) anchored in understandings of the bike and the process and activity of cycling in community. First, I explain how biking saved my life and how this moment was anchored in nature; the role of the bike as a technological disruptor; how the bike fueled dreaming and liberation, and the significant role and "covering" of community. Next, I map a freedom scie-cycling (FScF) framework, a nesting for finding home in the academy. This FScF engages a discussion of four tenets—Momentum and Balance, To Draft or to Lead, Justice in Motion, and Liberating Imagination—anchored in the science of the bike, biking, biking in the community, and the role of teamwork in group and competitive professional cycling. While my story is bike and biking centered, the FScF provides a trail map to guide other scholars to identify and understand their spaces and places, and counterspace(s) that is a home—to feel and be free and thrive, to anchor joy, and thus foster imagination and dreaming.

12.1 How Biking Saved My Life: An Anchoring in Nature

12.1.1 If My Helmet Could Speak

By Noemi Waight

If my helmet told you the stories,
Of the many miles of the hum yet, silence.
Of the release of pressure, of the letting go.
Of the wind caressing my face, allowing for the touch of the wind, reawakening feeling.

Of letting go of emotions, raw, rabid.
Of streaming tears, of water droplets, that cast off worry, fear, disappointment, struggles.
As quickly as they are released, evaporated into the atmosphere.
Becoming clouds that in turn become blessings of moisture.
Turning attention to the trees, grass, lake, colors, smells, butterflies, birds chirping.
To relaxation, almost as if unwinding, muscles adjusting to movements.
Motion becoming cyclical and morphing into Zen.
Reoccurring cyclical patterns of pace, thought, unthinking.
Exposed, yet covered.
Miles of opportunity and possibility.
Of joy, of contentment.
My helmet has seen, felt, listened, avoided, savored, remembered.

Miles (2023), in her historical account of *Wild Girls*, tells the story of how "nature classrooms taught these girls how to question societal beliefs and practices—and sometimes how to evade those who sought to steal life" (p. 40). In charting the history of Harriet Tubman, Harriet Jacobs, and Laura Haviland, Miles identified a consistent pattern where she noted that these women who changed the United States of America, shared one commonality, which involved cultivating and understanding their social worlds in relation to their natural surroundings. For example, Miles tells us the story of how the sun represented "hope, optimism, and ultimately freedom," where "the woods were places of refuge and spiritual rejuvenation," and bodies of water "were routes of escape, sources of harm or both" (p. 36). In other words, nature functioned as healing and a path to safety and liberation. For Harriet Tubman, the forest was a place of shelter that allowed her to breathe, dream, and navigate. In this account, Miles helps us understand how the forest and nature were simultaneously a place of resistance and hope for Harriet Tubman, Harriet Jacobs, and Laura Haviland.

In the spirit of the power of nature to heal and console—the bike, which shares an enduring history as a technology of resistance and liberation—functioned as a counterspace and medium to find home. As a child, I learned to ride my bike by propelling from one stilt (post) to another. In Belize, wooden homes were traditionally built on stilts (posts) that served a load- bearing function, and for me, served as guideposts that anchored as supports, as I rolled from one to the other, propelling my momentum and learning how to balance at the same time; after doing this a few times, my equilibrium stabilized, and soon I was rolling unassisted. At the time, this act alone functioned as an accomplishment and one that later allowed me to traverse more distance. As Kraft (1978) reflected on his process of learning to ride a bike, he noted a universal learning pattern that is defined by context and personal experience; we learn to ride a bike through experience from failure (e.g., falling) and from success (e.g., experiencing equilibrium and using force). This notion of learning brings to the fore engagement with self, the medium, and the process. This wobbly start (Glaskin, 2019) often gives way to mastery and confidence—the process of how to ride a bike recedes into the background as it becomes a second nature activity.

For me, this confidence morphed into play where I, my siblings, and neighborhood friends would spend countless hours riding up and down on our street, and as an extension, it became a mechanism to run errands for my mom.

At the height of my tenure process, I bought a bike from craigslist and a year later, I acquired a second bike that I dubbed the "gift that keeps on giving" (my family and friends contributed to this gift). Immediately, biking became an almost daily experience and soon, the bike would profoundly re-direct my professional and personal trajectory. At this juncture in my academic career, the bike re-emerged as a reminder of carefree, fun childhood and youth memories, and as a transformative medium for imagination and possibility.

12.2 The Bike

Turpin (2018) notes that "the bicycle has long been a part of American [U.S.] culture, but its image and significance have been ever changing" (p. 1). The bicycle has a long, enduring history. Analysis of this history reveals the bicycle as a tool that has been used for exclusion, segregation, elitism, forging subcultures, defiance, disruption, and liberation. Thus, it is no surprise that bikes, like other technologies, were once symbols of aristocracy, masculinity, and recreation. Bikes only moved into functionality and wider use after the design of the safety bicycle—a low wheeled vehicle with a diamond frame and a chain drive on the rear wheel" (Bijker, 1997, p. 2). What marked this transition from the earlier bike design was the incorporation of a mechanical means of propulsion, a chain drive on the rear wheel. In thinking about this history, Bijker (1997) poses an important question: Why did it take more than a half century for gears and a chain drive to appear on a working bicycle? The history of bicycles helps us understand the numerous iterations of design, which were informed by the sociocultural and political landscapes. Not surprisingly, the antecedents of the bicycle—or some iteration of it— was exclusively the domain of men and young boys. For example, the Boneshaker, which was very costly, was highly valued by men at Harvard and Yale.

It is not my goal here to chart the history of the bicycle; other writers have told this story in numerous forms (e.g., Bijker, 1997; Turpin, 2018, 2024; Wilson & Schmidt, 2020). Instead, what is important for the purposes of this chapter is to understand how the evolution of the bike and its use and adaptation has undergone numerous transformations based on the users and need. That the bike is symbolic of my own healing, and as a form of resistance and disruption of harmful and egregious practices, is significant. For almost 10 years, I was the only Black scholar hire in my department and STEM for that matter. Both my pre- and post-tenure experiences were laden with racism, gatekeeping, retaliation, exclusion, and attempts to silence, among other harms. Perhaps one of the most egregious experiences was the lack of acknowledgement and celebration (a precedent from previous years that publicly celebrated other colleagues) after earning tenure and promotion. Even after working twice as hard, publishing in top journals in my field, exerting tremendous sacrifice,

I was explicitly told that my accomplishments could not be celebrated because my White colleagues should also be successful: "It is [a] very difficult time for the department, something we have never faced before-we have three cases with negative votes in PRB in Advocacy. As I told you in person, I have NOT announced your tenure in public because it would naturally bring up the other three cases—which are all in danger of turning out badly" (email communication). This experience broke me—it robbed me of the joy and celebration of the moment. However, it did not destroy me and instead, it fueled my core and my purpose. That the bike and biking became a counterspace to manifest joy and liberation was a critical time and juncture in finding home as an academic scholar. While I write here about how the bike fueled my own liberation to counteract harm, it is important not to romanticize and omit how this technology historically was manipulated to exclude, discriminate, separate, and promote hate; it served as a mechanism to uphold racism, sexism, and ableism.

While I did not deliberately venture to examine the bike as a technological marvel that has endured for more than 200 years, my formative contributions—understandings of the nature of technology, the history and culture of technologies, and equitable, socially just technologies (see Waight & Abd-El-Khalick, 2012; Waight et al., 2022), framed my re-seeing of the bike as a technology and thus emerged the current focus of my scholarship. That the bike and process of biking is a STEM manifestation, a human-centered technology, and rooted in community, the birth of the STEM Cyclists program was a natural outcome. The bike and biking became this breathing synergy of liberation, opportunity, dreaming, and imagination. That robust rooting that I noted above would soon bloom.

12.3 Of Dreaming, Of Freedom

> I stepped away from the noise because I needed to think. And my happy place to do this: on a bike ride. And look at that sky! You see, our ideas come from the sky, the waves and calmness of the lake, from that gentle wind, from bug proteins, from the blue abyss, from the stillness, and from the sun rays peering through. We are scholars that build from the ground up, from nuggets, from that idea to a fully-executed project. For me, this has all happened on the bike.
>
> - Noemi Waight

Kelley (2022) writes about his mother and her desire for her children to find peace and see beauty and possibilities in the most mundane and fragile of their surroundings:

> My mother taught us that the marvelous was free...She simply wanted us to live through our third eyes, to see life as possibility…She wanted us to visualize a more expansive, fluid, cosmos-politan definition of Blackness, to teach us that we are not merely inheritors of a culture but its makers. (p. 2)

As Kelley so vividly describes, this wanting on the part of his mother arose from her own inner peace—what Kelley highlights as bliss—his mother's dreaming, which held visions of change and being free. It is telling that for Kelley, this manifestation of peace so aptly guided his own pathway, where rather than expending energy in anger, frustration, and struggle, he leaned into his imagination of a different future, a future of possibilities. In my own pursuits, I quickly learned that oppressive spaces and places are even more malicious and cruel when you understand that these actions are intended to rob us of our peace, joy, and our opportunities to dream and be free. After being retaliated for truth telling and refusing to participate in equity performance, and which resulted in being denied merit salary (and being told I was denied because I was not as good enough as my White colleagues who received merit pay [documents that report on how discretionary funds are allocated, are public], I quickly learned that the institution would not protect me and instead, inaction was the norm. Spending time telling and re-telling and hoping for action (and there was none), inflicted further harm. Kelley reminds us that to engage in struggle is to be constantly destabilized—in other words, the energy it takes to fight for parity robs one of the natural equilibria, and it requires even more energy to re-calibrate, to be productive, and be creative.

It took several years of my pre-tenure and even post-tenure years to acknowledge that our gatekeepers—those who spend their time retaliating and erecting barriers, of moving gold posts to stagnate, and in some cases destroy the career of scholars of color, do so because of their own shortsighted, chaotic, and mediocre accomplishments. They act to fulfill racist and White supremacists' expectations. You see, the act of dreaming, of imagining and creating, of building and moving, of shifting and transforming, of bridging and connecting, of making and re-mixing, are not acts that come from privilege, of being accorded visibility by virtue of privilege, of being named, recognized, and celebrated, and of knowing that you are not "the only one" in the room, program, office, or leadership. Acts of dreaming, of being free, of creating and making, are those of being "the only one," the singular or minoritized voice of the oppressed, of the gatekept, of those hoping for a better future, for transparency, for transformation, for peace-being, for joy, and for the simple act of being humanized. The roots of change and transformation come from acts and places of resistance (not from access accorded via White privilege) and are made first in dreams.

My bike and bike rides were my counterspace, an opportunity to feel and let go, to think, to create, to re-see, and to re-envision an outcome that I knew was doable—because I had done it—that is, written the papers in top journals, conceptualized, planned, and enacted research projects and programs. Being on the bike was dynamic, in constant motion, and importantly centered. It was these moments that gave me momentum and served as the impetus to re-imagine opportunities for self and community liberation. Even though so many of these bike rides traversed the same trails, each time new natural beauty was revealed. Time and again, the wildflowers, the crashing waves, the ripples on the lake, the bugs, and the gentle breeze provided new opportunities to re-see, revise, re-do, re-calculate, and re-experience possibility. Momentum on a bike is maintained through constant movement, and it fades through conscious and unconscious thought. It is being both present and absent, engaged, and mindless. Some bike rides were about being lost in the moment, with no destination in mind, while others required attention and provoked real thought. Regardless of one's state of mind, there was a constant: momentum, hence balance.

Kelley (2022) centers the main focus of *Freedom Dreams* on conversations about "what kind of world we want to struggle for" (p. 7). With this question, Kelley re-directs us away from simply existing in a milieu, to re-create how we want to exist. In parallel, my bike rides brought into sharp focus what is important, and it crystallized the essence of balance, and not if we move, but how we "move." When movement fosters re-engagement with self, when it provokes innovation and healing, when it fosters a joyful experience, then I argue, that is meaningful and productive movement. Kelley reinforced that it is not only what we fight against, but what do we fight for, and what do we dream for? That we fight for visibility, for movement from the margins to the center, for respect, and for our humanity. Interestingly, what I dreamt for was not related to the latter but rather, for joy, peace, and the ability to impact Black and Brown communities via the STEMcyclists program. Revolutionary dreams are born of political engagement, defiance, and disruption of the status quo. It is in these spaces and moments that we buck tradition and where revolutionary dreams are catalysts for new knowledge, possibility, and innovation. My counterspace fostered this possibility—it was the impetus for incubating ideas. Moments of peace fostered on the bike fueled creativity, and it was on this journey that the STEMcyclists—the science and engineering of bikes and biking—program was rooted.

12.4 Community in Motion

> We can imagine the future philosopher [Grace Lee Boggs] on these humid, summer days, pushing the pedals of her bicycle while she thought and dreamed in motion. (Miles, 2023, p. 115)

Miles (2023) tells the story of Boggs and how the outdoors nurtured her spirit and fueled her agency and activism. It was on vacation on the Long Island Southshore that Boggs rode her bicycle, which marked women's (though not all women's) independence in the 1890s and 1900s. The bicycle was thus a symbol of independence and freedom dreaming. Grossman, Wineburg, and Woolworth (2020) differentiate between community and society and specifically underscore how community is about cultivating "stronger bonds of connectedness" (p. 8). Thus, there is a sense of care with meaningful and loyal relationships. Community is defined by a stable, social structure that is supportive and is supported. Community is grounding, stability, and simultaneously dynamic and complex. By extension, community also involves personal responsibility—a commitment to and intentional engagement that promotes support and is loyal to the people, places, and spaces where one resides.

In essence, community is partnership and shared values. Grossman and colleagues (2020) explain that the contrast of community is a culture of individualism that breeds isolation and a lack of connection, a loss of community which defaults to individualism and isolation. For scholars of color in Predominantly White Institutions (PWIs), supportive community is often elusive. Generally, the burden to build community is often left to people of color, who exist as one of, or among a few in departments, schools, and colleges in the academy. In contrast, culturally, the bedrock and anchoring for people of color is family, friends, and extended communities. For many, the academy's cultural competence of people of color and community is generally deficient because it lacks knowledge of, understanding, and know-how of authentic community connections. Grossman et al. remind us that true community encompasses social interactions that prioritize "we" and "us" above the "me" and "I."

I found community in "we" and "us", in genuine connections shared via smiles, nods, a simple hello, an acknowledgement of "I see you"—of visibility, of value. I found community in the streets of Buffalo, among strangers who heard my accent and knew I was not of the born soil, all at the same time, saw and felt my humanity, saw and felt the welling of sadness in my eyes; they understood my yearning for stability, for roots, for connection, and for community. I found community in a quiet embrace, a nudging of the heart, in welcome, in acceptance, in belonging. Community became my anchor, and the bike and bike riding became the mediator for my community. It was off the beaten path that I became connected to places and spaces that were removed from the institution. The distance and serenity buffered the harm, the gatekeeping, the racism and White supremacy, the power-driven egos and personalities. Community formed an encasing, an armor that kept harm hovering and at the periphery but importantly, it negated entrance into sacred spaces and places. Grossman et al. (2020) reinforced how it takes time to cultivate safe vibrant and engaged communities that are anchored in trust and meaningful connection. The safety of community is critical for people of color who are transplants to new communities. People often seek safety of and connection with community in churches and religious affiliation, in sisterhood and brotherhood organizations, in professional spaces, and among family and friends.

I arrived at the Martin Luther King, Jr. Park in Buffalo, NY, the site of my first group bike ride, alone; hosted by Slow Roll Buffalo, and as I looked around, there were no familiar faces in sight. Nervous and with trepidation, I rolled my bike up to the main group of riders where I sought the sidelines. I looked around and marveled at the very large group of bike riders who had assembled to ride and be in community with one another. As I waited for the ride to start, a gentleman introduced himself and noted that he made it a point to meet and talk to someone new at every bike ride. It was as if he sensed my awkwardness, my sense of un-comfortability. While I never saw that gentleman again, unknowingly, that encounter influenced how I experienced my first group ride. Group rides became the norm, and I started showing up wherever rides were advertised. Eventually, strange faces became familiar faces, smiles, greetings, and embraces. It was on these rides that I started learning about the neighborhoods, community, history, and culture of the city of Buffalo. Every corner offered a new (to me) revealing history, shifting energy, fragrance, flavor, and re-connection with the humanity of people and people of color. It was on bike rides that I learned about past and present communities, transitions and shifting demographics, histories, and culture. It was this new knowledge and re-seeing that anchored my connection to Buffalo. It is important to note that before my first group ride, I had spent several years riding on my own—not alone. I was always in the company of my shadow, the rustling wind, and the nods of strangers. Beauty emerged and re-emerged on the bike path. Semblances of normalcy occurred on the bike away from the academic hallways, and it came just in time.

12.5 Freedom Scie-Cycling: Home as a Counterspace

These ordinary, yet extraordinary bike rides fueled joy, wonder, and curiosity and formed the core of how I now live and connect to self, of how I think of and do my work in the academy. That these bike rides were the catalyst for my personal, intellectual, and communal growth, imagination, and liberation, is significant. Traditionally, most of the research related to faculty of color in PWIs focus on systems of oppression, challenges, and struggle. While the latter is important documentation, it is simultaneously critical to document the stories and counterstories of scholars of color who made a way, made their own way, made their way with support from mentors, sponsors, friends, and family to thrive, be creative, and be innovative, to cultivate joy and be joy. In defining notions of joy, Okello (2024) extends the work of Tichavakunda (2021) to conceptualize Black joy as unspeakable, and emphasizes, "Black flesh, and its interactions with various land and soundscapes, as an important site of being" (p. 8). That joy was a key factor of finding home on the bike, across spaces and places anchored in the tranquility of, and jazzed sounds of nature is consistent with how scholars of color have pushed back on harm within academic hallways. To be joy and summon joy is a state of being, an intentional practice that counter attempts to deny, erase, and silence our voices and personhood.

For scholars of color, to be joyful is to cultivate dispositions, practices, and ways of life that resist oppression. Mentors, academic sponsors, family, friends, and community spin a cocoon of protection, helping to support and nurture strong roots. This encasing, which is temporary, functions as a scaffold, eventually allowing for a "release" that is stronger and more robust. This robustness is anchored in the soul, in the knowledge of our ancestral, familial, and cultural heritage. This is not to suggest that strong winds, pests, and other eco-environmental challenges are eliminated, but rather that the anchoring and purpose, that the possibility and hope are hardier.

In my own trajectory, I have been fortunate to have consistent and culturally responsive mentoring and support from my advisor and other prolific scholars, colleagues, community members, and most importantly, my friends and family. Here I think of opportunities that have given me platforms for scholarly visibility, international support and collaborative efforts, access to communities such as Continental and Diasporic Africa in Science Education (CADASE) at NARST (a global organization for improving science education through research), among others, and prolific role model scholars such as Dr. Mary Atwater. These stories are essential to serve as mirrors and models of women of color who have authored their own trails, who have re-purposed alternative routes for productive and joyful careers. As people of color, we are at our best, most productive selves when we nurture our physical and mental well-being. In her talk for CADASE during the 2024 NARST conference,

Dr. Felicia Moore Mensah contrasted well-being with being well (Moore Mensah, 2024). Ultimately, being well is marked by numerous facets of our professional and personal lives. Importantly, when our well-being is optimal, we are freer to dream, take risks, and seek freedom.

In my journey, the bike—as a transparent technology and as a tool for resistance, disruption, joy, and liberation—was the medium to engage freedom scie-cycling. Below, I engage a discussion of a Freedom Scie-cycling framework (FScF) anchored in understandings of the bike and biking as a space and an activity of cycling in community. I argue that the bike as a tool and mechanism for disruption, freedom, and dreaming serves as an activity and counterspace for liberatory practices and joy-filling. In the next section, I promote four tenets that inform my conceptualization of FScF: Momentum and Balance, To Draft or to Lead, Justice in Motion, and Liberating Imagination.

12.5.1 Momentum and Balance

The early models of the bike saw various iterations that relied on the feet for motion. For example, the running machine was moved forward by pushing on the ground with the feet (Bijker, 1997). In subsequent designs, the goal was to raise the feet from the ground. This effort saw various iterations of the velocipede, which eventually gave rise to models that informed the current standard. The emergence of the air-filled tire was critical in shaping new designs and functions of the bicycle. Fast forward to current models, it is well understood that certain features such as triangles that make up the diamond frame distribute forces and inform bike stability. In addition to structures like the frame, the momentum of the rider is also important in maintaining the bike upright.

There is a common saying: to stay in balance, you must keep moving. This is the principle of momentum that allows the bike to stay upright. In my own journey, balance and stability came from momentum and consistency. How balance shows up in our professional and personal lives is significant for how we experience the academy. As we contemplate our identities as scholars, are we only professors and doctors? Does this identity define us as individuals and our personhood? If this is the case, it suggests that our balance is tipped, and this can be harmful for our spirit and psyche. Indeed, Moffit et al. (2012) reported how women of color, despite their marginalization, stood in their empowerment because of their identities as women of color. My identity as a woman of color, a person who is differently abled, a cyclist, a league cycling instructor, an innovator, a scholar, and a professor communicates that my work exists at the intersections of numerous life experiences. To suggest that I should only be defined by a singular identity would mean betraying the very core of who I am meant to be. Thus, how we cultivate, and which identities we cultivate contributes toward balance and stability. While momentum on the bike keeps us balanced and upright, we are always navigating multiple forces—air resistance, friction, and gravity—at the same time. However, the key here is to stay in motion. For sure, there are numerous factors that impact the type of motion, but what is most important here is balance.

12.5.2 To Draft or to Lead

Glaskin (2019) explains drafting this way:

> there is one way of harnessing the effects of aerodynamics that doesn't cost anything—riding in the slipstream of another cyclist. Put simply it means the rider in front does all the hard work of pushing the air aside while the following rider sits behind. (p. 148).

Similarly, Hirsch and Levy (2013) state that "drafting involves a group of cyclists moving together… the phenomenon of drafting offers unique insights into the delicate balance between competition and cooperation among collectives" (p. 10). In cycling, teamwork is essential to engage a safe and efficient ride. An important facet of group riding involves the benefits of both leading and drafting. To lead a group ride is to be a leader, decision maker, a buffer, a visual, a protector, and the pusher of air. To draft is to be protected, to ride in the shadows and behind the leaders and other riders who mitigate the effects of elements such as wind and friction for a smoother, efficient ride. Scientific research provides evidence via oxygen consumption that those who ride as the third in line or in the middle of a peloton do half the work and are less tired when compared with the leader and those in front of the ride (Glaskin, 2019). Naturally, Glaskins explains that it is easier to follow than to lead. However, in cycling, unlike other activities, every rider takes a turn to lead and to bask in the buffer of the leader and front riders. In my early career, I led and continue to lead most of the manuscript writing, and in the process offering support for colleagues and graduate students. In this way, every rider understands the multiple and important roles that cyclists undertake to support each other. Usually, based on the order on the ride, there is a shifting interplay of technique, skill, and knowledge. That cycling offers this opportunity to shift roles as opposed to sitting statically in singular roles is critical to our growth. To experience both roles is to have a unique perspective, but importantly to develop a balanced understanding of and appreciation of the varying roles involved in cycling. These roles expand even further when the rides are in large groups—roles such as corker, sweep and sag—are key in ensuring safety and a fun community ride.

Cycling thus provides a critical framework, supported by scientific research, to organize how we dream and imagine more justice-centered possibilities and futures. To draft is to experience support and feel supported. It is well documented how supportive spaces and communities—mentoring and specifically culturally relevant mentoring—impact our productivity and how we thrive. Cultivating a village of support and by example, learning to support others, is the foundation of culturally relevant care (Watson et al., 2016). The shift from cycle team leader to drafting is one that involves care and sacrifice. As Hirsh and Levy (2013) explained, it is a delicate balance between competition and cooperation. There is a natural reciprocity of being a giver and a receiver, which involves sacrificing so that those in the pack can come out ahead. Fundamentally, trust and loyalty buffer this collective endeavor. Biking provided this cover, a protection that allowed me to regain my sense of self, to re-store my body and mind, to re-center my core, re-shape my psyche, and re-cover my value. With group biking, I often thought of riding in the middle as protective space—an opportunity to exhale, relax and build endurance so that I could go further and faster. When cycling in groups, there is an inherent understanding and trust that all riders look out for each other, be aware and cognizant of traffic, other cyclists, walkers, runners, and road hazards. Community rooted in trust and loyalty serves as momentum to leap, take risks, and dream. When we know someone has our front-and-back, it defines our feeling of security and safe spaces and places, and thus we are more likely to take risks, to be innovative, and dream. When our sense of balance is re-stored, we restory narratives of possibility—effectively, the bike was a source of momentum, for liberation, to engage imagination, and to be with and in community.

12.5.3 Justice in Motion

> … we will shine a light on those who may at times feel crushed, but who crawl out from under the social structures bearing down on us, find others who want to do more than survive, and whisper to one another— 'They tried to bury us, they didn't know we were seeds.' (Benjamin, 2024, p. 77)

The bicycle has been used to disrupt the status quo. Simultaneously, the bicycle was used to perpetuate racism, to exclude, disenfranchise, and segregate people of color, primarily Black people. From 1894 through the 1990s, The League of American Bicyclists actively banned Black people from joining the league. They essentially promoted a White organization that excluded Black people from participating in cycling sports and bicycle-related events and programming. Effectively, rendering Black people invisible in the cycling community resulted in the perception that biking was a White activity, and thus for a long time, this lack of representation omitted the significant contributions of Black Americans in cycling in the United States. That the actions of the League of American Bicyclists reflected broader social and institutional systems of White supremacy is no surprise. Today, I am among a small cadre of people of color, who are certified League of American Bicyclists safe cycling instructors, beneficiaries of the league's attempt to move past their racist history. In many ways, our presence in the league and riding in community (like academia) is an act of defiance—a visible political act in motion.

Dynamic acts of justice counter our static existence, which moves us from struggle to flourish. Benjamin (2024) explains how power structures benefit from our lack of equilibrium which keeps us in a constant cycle of struggle; it is akin to our psyche being "arrested and confined, making it difficult to think beyond the racist, classist, sexist, ableist status quo" (p. 75). To exist in constant struggle is to repress our imaginaries and possibility—for us in the academic enterprise this means stifling our creativity, ideas, and words that constitute our writing. If we don't write, if we are not creative, then we don't innovate… and by extension, we give institutions "permission" to gatekeep our progression and promotion. In her discussion of imagining justice, Benjamin (2024) is explicit that imageries of possibility are ounterspaces to defy "those who refuse to accept oppressive ideologies" (p. 76). Here, it is important to highlight that while the goal is freedom from the racist and oppressive gaze, this effort is no small feat. It is a revolutionary act to find and nurture that which heals and feeds our soul. Justice in motion is always in flux and in constant interaction with the numerous forces that slow us down, halt our motion, but equally, can propel us to move faster and further. When our cadence stabilizes and our bodies relax, we can tune out the noise and welcome silence, peacefulness, and joy. This state of being is marked by possibility and dreaming.

12.5.4 Liberating Imagination

To think of imagination is to bring to the fore the possible and the impossible. Benjamin (2024) notes that her definition of imagination is unruly and is about coloring outside the lines: "my use of the word is undisciplined, promiscuous and porous" (Preface). She explains how she shifts between imagination and imaginaries and interchangeably thinks of imagination as dreaming. Indeed, to live life on the bike and to use the bike as a space for resistance is to ride off with no set plan and no perfect destination. What results is a potpourri of possibilities and re-newed experiences. Riding on the lakefront hundreds of times, yet experiencing the lake, the sky, the trail, the seasonal changes, and the people differently every single time. To color outside of the lines while biking is to be oneself, to do things on your own terms. While the eight hundred and thirty-three (877) rides—an equivalent of 7,801 riding miles that I have recorded over time—this distance and time is miniscule when compared with lifetime achievements of professional and seasoned cyclists. What I celebrate here is consistency and commitment of short rides versus long rides and speed. These rides have retraced similar routes, streets, neighborhoods, and communities. Each time, there is either a reinforcing of the cultural, historical, and social history or the emergence of new knowledge. Coloring outside the lines while biking involves forging new pathways, defying expectations of who can and should ride, transgressing boundaries, and being the only one or among a few people of color and/or the only differently abled person of color. Coloring outside the lines in academia involved standing apart or alone, speaking up even when it was not safe or the popular thing to do; it involved daring to dream, conceptualize, and build innovative projects.

Bike riding became my safe space, my opportunity to reflect, to rest and de-stress, to temporarily pause institutional harm, and to summon joy. This was my opportunity to live and to defy systems of power that sought to keep people of color in loops of subservience. In response, my loops were transformed into loops of possibilities and dreams. Indeed, Benjamin (2024) explained that to engage imagination and to dream is to counter and put in their place "those in power who work tirelessly to squash us from having radical imaginations that dare to envision a world in which everyone can thrive" (preface).

In this context, Benjamin (2024) engages imagination of self, justice, future, and incubation. Here, Benjamin (2024) cites Perry, noting how "imagination has always been our gift." In her reflections of the Pelican program, what she classified as "Freedom Fridays," she recalls an activity that guided fifth graders to think of new uses for ordinary objects like scissors, rubber bands, and so on. She noted that it was these kinds of meaning-making activities that fostered divergent thought processes. It was these kinds of activities that were catalysts for innovation and invention. Yet, Benjamin recalled that this opportunity was limited to very few students of color. It begged the question of why these opportunities for divergent and critical thinking were limited and continue to be reserved for only a select few students. Benjamin argued that all students should have opportunities for liberating imagination—radical opportunities for action that transforms "our identities, our psyches, our spirits" (p. 43).

On these bike rides, I dared to dream. In his analysis of unspeakable joy Okello (2024) thinks with Harriett Jacobs and reifies that "imagining otherwise, as noted above, is a desire to live the future we must live in the now, present moment" (p. 19). He stresses that dreaming and the imagery of dreams is not often "sayeable" and thus these imageries and visions function as potent potentialities. Freedom Scie-cycling as home—a space, an activity and process—functioned as a safe and soft landing that bolstered dreaming, and liberatory joy. So when challenges came my way, when gatekeepers raised their standards (for scholars of color), when silencing and erasure were used as mechanisms to disenfranchise, when resources and opportunities for wealth building were denied… every single time, my bike ride offered time to think, to strategize, to process, to dream and understand that these moves were not of mine and that our most potent potentialities resided in us, in me, in our brilliance, in our freedom, in our dreaming, in our joy, and in love. These bike rides were not only responses to challenge, but also of celebration, of small wins, of moments of beauty, of smiles and laughter, of dreaming, of opportunities. Some bike rides were celebration, joy, unspoken glee, and love. From an academic standpoint, the small wins were research outputs: manuscripts, grant proposals, national and international research collaborations, science education teacher professional development, and planning, designing, new door openings, and bringing to life the STEMcyclists program. From a personal angle, the wins involved community connections, community building, nature connections, forging humanizing connections, blazing new trails, trust, and friendships, and love. Altogether, bikes and biking as a counterspace and as a new home converged pedagogy, research, community, nature, humanity, joy, liberating imagination, and love.

References

Benjamin, R. (2019). *Race after technology: Abolitionist tools for the new Jim code.* Polity Press.

Benjamin, R. (2024). *Imagination: A manifesto.* W.W. Norton & Company Inc.

Bijker, W. E. (1997). *Of bicycles, bakelites, and bulbs: Toward a theory of sociotechnical change.* MIT Press.

Glaskin, M. (2019). *Cycling science: How Rider and Machine Work Together.* Ivy Press.

Grossman, P., Wineburg, S., & Woolworth, S. (2020). *What makes teacher community different from a gathering of teachers.* University of Washington.

Hirsh, A., & Levy, S. T. (2013). Biking with particles: Junior triathletes' learning about drafting through exploring agent-based models and inventing new tactics. *Technology, Knowledge and Learning, 18*(1–2), 9–37. https://doi.org/10.1007/s10758-013-9199-8

Kelley, R. D. G. (2022). *Freedom dreams: The black radical imagination.* Beacon Press.

Kraft, R. G. (1978). Bike riding and the art of learning. *Change, 10*, 36–42.

Miles, T. (2023). *Wild girls: How the outdoors shaped the women who challenged a nation.* W. W. Norton & Company.

Moffit, K. R., Harris, H. E., & Forbes Berthoud, D. A. (2012). Present and unequal: A third-wave approach to voice parallel experiences in managing oppression and bias in the academy. In G. G. y Muhs, Y. F. Niemann, C. G. Gonzälez, & A. P. Harris (Eds.), *Presumed incompetent: The intersections of race and class for women in academia* (pp. 78–92). University Press of Colorado. https://doi.org/10.2307/j.ctt4cgr3k.12

Moore Mensah, F. (March, 2024). Science education for the rest of us: Unifying our community for our well being and for the rest of us. Keynote for the CADASE RIG: NARST International Conference. Denver, CO.

Niemann, Y. F. (2012). Lessons from the experiences of women of color working in academia. In Y. F. Niemann, G. G. y Muhs, C. G. Gonzälez, & A. P. Harris (Eds.), *Presumed incompetent: The intersections of race and class for women in academia* (pp. 446–500). University Press of Colorado. https://doi.org/10.2307/j.ctt4cgr3k.40.

Okello, W. G. (2024). Unspeakable joy: Anti-black constraint loopholes of retreat, and the practice of Black joy. Urban Education, 1–28. https://doi.org/10.1177/00420859241227962

Tichavakunda, A. A. (2021). Black joy on white campuses: Exploring black students' recreation and celebration at a historically white institution. *The Review of Higher Education, 44*(3), 297–324. https://doi.org/10.1353/rhe.2021.0003

Turpin, R. J. (2024). *Black cyclists: The race for inclusion.* The University of Illinois Press.

Turpin, R. J. (2018). *First taste of freedom: A cultural history of bicycle marketing in the United States.* Syracuse University Press.

Waight, N., Kayumova, S., Tripp, J., & Achilova, F. (2022). Towards equitable, social justice criticality: Re-constructing the "Black" box and making it transparent for the future of science and technology in science education. *Science & Education, 31*, 1493–1515.

Waight, N., & Abd-El-Khalick, F. (2012). Nature of technology: Implications for design, development, and enactment of technological tools in school science classrooms. *International Journal of Science Education, 34*, 2875–2905.

Watson, W., Sealey-Ruiz, Y., & Jackson, L. (2016). Daring to care: The role of culturally relevant care in mentoring Black and Latino male high school students. *Race, Ethnicity and Education, 19*, 980–1002.

Wilson, D. G., & Schmidt, T. (2020). *Bicycling science.* The MIT Press.

Index

M. M. Atwater and B. A. Garvin (eds.), *Creating and Maintaining a Home in Science Education Research*, Contemporary Trends and Issues in Science Education 65, https://doi.org/10.1007/978-3-031-99382-4

MIX
Papier aus verantwortungsvollen Quellen
Paper from responsible sources
FSC® C105338

If you have any concerns about our products,
you can contact us on
ProductSafety@springernature.com

In case Publisher is established outside the EU,
the EU authorized representative is:
Springer Nature Customer Service Center GmbH
Europaplatz 3, 69115 Heidelberg, Germany

Printed by Libri Plureos GmbH
in Hamburg, Germany